高等职业技术教育系列教材

航空工程专业英语

主　编　陈凯君　丁镜之　王　鹏

副主编　邵绪威　夏罗生　赵翔鹏

　　　　唐道湘　李亚非

主　审　邓青霞

西安电子科技大学出版社

内 容 简 介

　　本书是供航空类高等院校学生学习航空工程专业英语的教材。全书共21课，由三大模块组成。模块一为航空概论，给出了航天、航空的定义，阐述了飞行原理。模块二为飞行器，详细阐述了飞行器方面的知识，包括飞机部件、电源系统、液压系统、气动系统、燃料、飞行导航、飞行控制、起落架等内容。模块三为燃气涡轮发动机，详细阐述了燃气涡轮发动机方面的知识，包括燃气涡轮发动机进气道、压气机、燃烧室、涡轮、排气装置、反推力装置、制造材料等内容。

　　本书可以作为本科或高职高专院校航空工程专业的英语教材。

图书在版编目(CIP)数据

航空工程专业英语/陈凯君，丁镜之，王鹏主编. —西安：西安电子科技大学出版社，2022.8
ISBN 978 - 7 - 5606 - 6532 - 0

Ⅰ. ①航…　Ⅱ. ①陈…②丁…③王…　Ⅲ. ①航空工程—英语—高等学校—教材　Ⅳ. ①V2

中国版本图书馆 CIP 数据核字(2022)第 119376 号

策　　划　杨丕勇
责任编辑　杨丕勇
出版发行　西安电子科技大学出版社(西安市太白南路2号)
电　　话　(029)88202421　88201467　　　邮　　编　710071
网　　址　www.xduph.com　　　　　　　电子邮箱　xdupfxb001@163.com
经　　销　新华书店
印刷单位　陕西日报社
版　　次　2022 年 8 月第 1 版　2022 年 8 月第 1 次印刷
开　　本　787 毫米×1092 毫米　1/16　印张　8
字　　数　162 千字
印　　数　1～3000 册
定　　价　22.00 元
ISBN 978 - 7 - 5606 - 6532 - 0 / V

XDUP 6834001 - 1
＊＊＊如有印装问题可调换＊＊＊

前　　言

近年来，我国的航空事业获得了持续、快速、健康的发展，航空领域对从业人员的专业英语阅读水平、翻译水平和交际能力提出了更高的要求。而目前大多数航空领域从业人员的专业英语综合水平有待提高，部分机务人员仅能简单地理解英文维修工卡，英语口头沟通和书面交流能力明显偏弱。

目前市场上适合航空院校使用的专业英语教材种类不多，且部分教材太过强调航空专业文章阅读，虽对提升英语阅读能力有一定的作用，但其实践性与岗位针对性不强，不能满足航空领域从业人员对英语综合能力的要求。有些院校节选飞机维修手册部分章节的资料作为教材，实用性虽强，却缺乏理论基础和系统性，更没有配套练习，使得专业英语教学达不到理想的效果。

本书根据航空领域从业人员的实际岗位需求，借鉴飞机维修手册的部分内容，再结合其他航空专业英语资料，配以实物插图编写而成。本书具有如下鲜明的特色：

（1）突出了航空专业特色。本书在编写过程中充分考虑了航空类专业交叉性、综合性强的特点，在要求学生掌握理论知识的同时，以培养综合素质高、能在航空企事业单位服务的复合型人才为目标。

（2）面向航空应用，注重实践能力的培养。适当拓宽专业基础知识的范围，以增强学生的适应性；面向航空工程实际，注重实践环节，强化在航空系统就业所必需的职业技能培养内容，以提高读者的实际动手能力和创新能力。

（3）对接岗位需求，体现"工学结合"。本书的大部分材料来自实际工作中使用到的维修资料，同时结合航空企业实际工作岗位的需求与工作程序增加了一些内容，有效地还原了工作岗位的实际需求。

（4）专业术语结合实物配图，图文并茂。本书将专业讲解与实物插图结合，图文并茂地展现了航空产品部件。通过学习本书，能方便地把抽象的英文术语与实物联系起来，从而使专业术语的学习更加直观和高效。

本书由张家界航空工业职业技术学院陈凯君、丁镜之、王鹏任主编，邵绪威、夏罗生、赵翔鹏、唐道湘、李亚非任副主编，熊志亮、谢志龙、刘榕、游泽宇参与了编写，邓青霞任主审。张家界航空工业职业技术学院航空维修学院刘让贤院长、凡进军副院长，航空制造学院胡细东院长、邵绪威副院长对本书给予了大量技术性支持和指导。在此，向对本书编写给予指导、支持、帮助的领导、同事表示真诚的感谢。本书参考了部分航空公司的书籍、资料和图片，在这里谨向相关作者及为本书提供帮助的机务维修一线的领导和技术人员表示诚挚的感谢。

本书可作为本科类和高职高专类航空院校的发动机维修、发动机装配调试技术、发动机制造、飞机维修、飞行器数字化制造、飞机机电设备维修、飞机电子设备维修、通用航空、无人机等专业的专业英语教材。

由于编者水平有限，加之时间仓促，书中难免存在疏漏之处，恳请广大读者和专家批评指正。

编　者

2022 年 5 月

目　　录

Section 1

Aviation

Lesson 1   Aeronautics Versus Aerospace

Aeronautics refers to the navigation of manned or unmanned aircraft through the earth's atmosphere; Aerospace refers to the navigation activities of manned or unmanned spacecraft outside the earth's atmosphere, also known as space flight or cosmic navigation (Figure 1. 1).

Figure 1. 1   Aeronautics vs. Aerospace

Before there can be a discussion about aeronautics, it must be defined a bit better. Aeronautics is the science of designing an airplane or other flying machine. There are four basic subjects that aeronautical engineers must understand:

(1) How to shape the airplane and its swing and tail so that it slips easily through the air and can lift itself off the ground. This is called aerodynamics.

(2) How to control the airplane so that it will be able to turn but not spin out of control. This is called stability and control.

(3) How to build an engine (be it jet or propeller) so that the airplane can push its way through the air. This is called propulsion.

(4) How to build the airplane so that it won't fall apart when it hits a gust of wind or slams down on an aircraft carrier. This is called structures.

Aeronautical engineering should not be confused with aerospace engineering, which deals with things such as rockets and satellites. While rockets must ascend through the atmosphere，they must also navigate through space where there is no air. Their designs and engines have very different requirements than those of airplanes.

Vocabulary

aeronautics	/ˌeərəˈnɔːtɪks/	n. 航空学；飞行术
navigation	/ˌnævɪˈɡeɪʃn/	n. 导航；航行，航海；航运，水上运输
aerospace	/ˈeərəʊspeɪs/	n. 航空航天工业，航空航天技术；航空与航天空间
		adj. 航空和航天(工业)的，宇航的
aerodynamics	/ˌeərəʊdaɪˈnæmɪks/	n. 空气动力(特性)；空气动力学
stability	/stəˈbɪləti/	n. 稳定(性)，稳固(性)；坚定，恒心
propulsion	/prəˈpʌlʃn/	n. 推进；推进力
structure	/stˈrʌktʃə/	n. 结构，构造；结构体，(尤指)大型建筑物；周密安排，精心组织；机构，组织，体系
be confused with …		与……混淆
rocket	/ˈrɑkɪt/	n. 火箭；火箭发动机
satellite	/ˈsætl. laɪt/	n. 卫星；人造卫星

Exercise

Complete the following sentences and translate them into Chinese.

(1) Aerospace refers to ＿＿＿＿＿＿＿＿＿＿＿＿＿＿＿＿＿＿＿ or unmanned spacecraft ＿＿＿＿＿＿＿＿＿＿＿＿＿＿＿＿＿＿＿＿, also known as space flight or cosmic navigation.

翻译：＿＿＿＿＿＿＿＿＿＿＿＿＿＿＿＿＿＿＿＿＿＿＿＿＿＿＿＿＿

＿＿＿＿＿＿＿＿＿＿＿＿＿＿＿＿＿＿＿＿＿＿＿＿＿＿＿＿＿＿＿

(2) Aeronautics is the science of ＿＿＿＿＿＿＿＿＿＿＿＿＿＿＿＿＿.
There are four basic subjects that aeronautical engineers must understand.

翻译：＿＿＿＿＿＿＿＿＿＿＿＿＿＿＿＿＿＿＿＿＿＿＿＿＿＿＿＿

＿＿＿＿＿＿＿＿＿＿＿＿＿＿＿＿＿＿＿＿＿＿＿＿＿＿＿＿＿＿＿

(3) How to control the airplane so that ＿＿＿＿＿＿＿＿＿＿＿＿＿. This is called stability and control.

翻译：＿＿＿＿＿＿＿＿＿＿＿＿＿＿＿＿＿＿＿＿＿＿＿＿＿＿＿＿

＿＿＿＿＿＿＿＿＿＿＿＿＿＿＿＿＿＿＿＿＿＿＿＿＿＿＿＿＿＿＿

(4) How to build the airplane so that _____

_____.

翻译: _____

_____

(5) Aeronautical engineering should not _____,

which deals with things such as rockets and satellites.

翻译: _____

_____

(6) While rockets must _____,

they must also navigate through space where there is no air.

翻译: _____

_____

+-+-+-+-+-+-+-+-+-+-+-+-+-+
Supplementary Reading
+-+-+-+-+-+-+-+-+-+-+-+-+-+

Perfect Gas Law

The perfect gas law establishes the relationship between the pressure, density and temperature of a gas at any instant in time or space. Even though it is a mixture of gases, air is mostly nitrogen and can be treated as a perfect gas. Engineers regularly use the perfect gas law to compute air flow properties. The equations for the perfect gas law were found empirically, meaning that many, many experiments were run and date compared mathematically to get the equations.

The simplest and most commonly used formula for the perfect gas law says that the pressure of a gas is equal to the density of the gas multiplied by the gas constant (a function of the atoms in the gas molecules) multiplied by the temperature of the gas. It is very useful in defining how the properties of a gas will change when one of the properties is changed. Fox example, if the density of a gas is held constant, but the container is heated to raise the temperature, the perfect gas law says that the pressure must also rise. A dangerous example of this principle is illustrated when an aerosol can is thrown into a fire. The density of this can is fixed; neither the mass inside the can or its volume can change. When the can heats up, it will explode when the pressure inside becomes too high for the can to contain!

Lesson 2   How Air Moves Over Objects

1. Different Ways Air Moves

The following definitions are terms that aerodynamicists use to define the way a fluid moved in or around an object. In order to get a good picture of what is happening about a wing, for example, the aerodynamicist must know the velocity of the plane, the altitude of the plane, the size and shape of the wing, and the properties of the air. He or she will use the terms and concepts discussed in this list to define the fluid flow.

2. Speed of Sound

Sound travels in invisible waves of changing pressure through a fluid (usually air, but sometimes liquid). If a person is standing very far from an explosion, he or she will not hear it right away. It takes time for the sound waves to travel. A person standing closer to the explosion will hear it sooner. At sea level, on a typical day (not too hot, not too cold), the speed of sound (how fast the sound waves travel) is 340 meters per sec (m/s), 1,115.54 feet per second (ft/s), or 760 miles per hour (m/h). The speed of sound is a function of the pressure and the density of the fluid in question. Since both the pressure and the density can change with temperature or altitude, the aerodynamicist must compute the speed of sound at the altitude, temperature, pressure, and density where the plane is flying.

3. Mach Number

Because aerodynamicists like to be able to compare results, they have defined a dimensionless (no units) value to measure the velocity of the airplane. The Mach number is defined as the velocity of the plane divided by the speed of sound computed for the airplane's altitude. An airplane at a low altitude flying at Mach 0.8 will have the same airflow behavior over the wing as the same airplane flying at a high altitude at Mach 0.8, the airplane at the higher altitude to be flying at Mach 0.8, its velocity will be slower than that of the plane flying at the lower altitude! The behavior of the airflow over the wing, however, will be the same on both planes.

The Mach number is named for Ernst Mach (1838 – 1916), who conducted the first meaningful experiments in supersonic flight while a professor of experimental physics at the University of Prague (Figure 2.1).

Figure 2. 1 Ernst Mach (1838 – 1916)

The Mach number is useful to aerodynamicists in another way, too. They use the Mach number to define characteristics of the flow and to determine the correct mathematical procedures to use to compute the flow behavior. Mach numbers less than 1. 0 define a flow regime that is called subsonic flow. Mach numbers greater than 1. 0 define a regime called supersonic flow. If the Mach number is greater than 5. 0, that regime is called hypersonic flow. Sonic conditions exist at Mach number equal to 1. 0. Because flows from about Mach 0. 75 to 1. 20 can have areas that are both subsonic and supersonic, aerodynamicists have named it the transonic regime. They must be very careful how they manipulate the fluid equations in this regime.

Another way to look at the flow regimes based on the Mach number is to study the waves generated by the plane flying through the air. In the subsonic, or relatively slow speed regime, the waves of changing pressure about the planes travel out in all directions at the speed of sound from that altitude.

As the plane flies faster and approaches the transonic regime (still below Mach 1. 0), the waves in front of the plane don't travel that much faster than the plane itself. The airplane's speed is approaching that of the speed of sound for the waves.

At the sonic barrier, Mach＝1. 0, the front of the waves and the plane are traveling at the same speed. The velocity of the plane and the speed of sound for the waves are equal.

As the plane files faster than the speed of sound (Mach number greater than 1. 0), the waves compress into a cone-shaped envelope about the plane. The flow conditions of the air ahead of the plane remain unchanged or constant until after the plane flies past. Only the region inside the cone is affected by the plane. This conical compression is called a shock wave, and it will be discussed in greater detail in a later section.

4. Friction

Anything that moves against another object must overcome the resistance to motion between the two objects. If a person tries to push a box across the floor, he or she must push hard to overcome the resistance. This resistance is a force called friction. If the person applies a push, or force that is stronger (larger) than the frictional force, the box will move. If the push isn't strong enough, the box won't move.

The friction between two moving objects can be affected by the surfaces of the objects. For example, it is easier to push a heavy box across a smooth-wood floor, or a sheet of ice, than it is to push it across a thick, bumpy carpet. That means the frictional force between the box and the smooth floor or ice sheet is less than the frictional force between the box and the thick carpet, so it takes less of a push to get it moving.

When a fluid like air flows across a surface such as a wing, there is friction resisting the motion. How much friction is dependent on mainly two factors, the viscosity of the fluid and the smoothness of the surface. A very viscous fluid like honey (a fluid with high viscosity) will resist flowing, even down a smooth surface. The friction force is very strong at the surface. A fluid like water with a much lower viscosity will travel much faster down a smooth surface; the frictional force between the water and the surface is much smaller. However, if water flows across a very rough surface, like carpet, it will travel down more slowly than on the smooth surface. Because the surface is rougher, the friction force is stronger, and the velocity is slower.

5. Boundary Layer

Because of this friction force, when a fluid flows over a surface, an interesting pattern develops. At the very surface, the fluid actually stops; there is no velocity or movement at the surface. But because the fluid is a deformable body, layers of the fluid above the surface move over the stopped flow. In addition, each layer experiences friction between it and the next layer, so each layer flows a little slower than the layer above it. Eventually, some distance away from the surface, there is no effect of the slowed flow, and the remaining layers of the fluid travel at the original velocity. This distance is called the boundary layer thickness, and all of the thickness all over the surface form the boundary layer.

In general, the boundary layer gets thicker as the flow moves along the surface. How fast and how big the boundary layer grows is a function of the smoothness of the surface, the shape of the surface, and how fast the flow is traveling.

6. Laminar Boundary Layer

For lower velocities, fluid flowing over a smooth surface that is relatively short and flat will only develop a very thin boundary layer.

The flow inside the boundary layer will be smooth and orderly, meaning that the layers will basically stay in layers, without mixing. This condition is called a laminar boundary layer.

Unfortunately, nature tends towards disorder, so it is rare to be able to maintain a laminar boundary layer for very long.

7. Turbulent Boundary Layer

As a fluid moves over a long, relatively flat surface, the boundary layer will get thicker, and the layers will start to mix and swirl around each other. This swirling, rolling layer is called a turbulent boundary layer. The mixing and swirling is called turbulence; if the swirling is regular and repeatable, it is called a vortex or an eddy.

Since mist of the boundary layers over an airplane will be turbulent, aerodynamicist will try to design the surfaces to minimize the amount of turbulence or disorder.

8. Transition

The region in the boundary layer where the orderly laminar layers start to mix together, but before they really start swirling, is called the transition region. Most of the time it is a fairly small region. In order to maintain control over the boundary layer, the aerodynamicist will often design "trip points" on the surface to trigger the transition to turbulent flow. They will try to design the surface to maintain a barely turbulent boundary layer.

9. Flow Separation

When a turbulent boundary layer really starts to swirl, the boundary layer thickness starts to grow even faster. Eventually the flow is so mixed, it starts to flow back towards the front of the surface! When this happens, the outside, original fluid is moving over a large bubble, and inside the bubble, the flow is moving back up the surface. This is called flow separation. The front of the bubble where the outside fluid turns sharply away from the surface is called the point of separation; the back of the bubble, where the outside fluid turns back to follow the surface again, is called the point of reattachment. If region of flow separation extends past the surface, this region is called a wake.

Pilots and engineers usually don't like it when the flow separates on a wing. This is a condition known as stall, and when a wing stalls, the lift (a force that helps a plane to fly, see later section) decreases sharply. The plane loses altitude, and if the stall is not corrected, the plane will crash. To land a plane, however, a pilot will wait until the plane is close to the ground, then initiate a slight, controlled stall to gently drop the plane to the runway.

10. Buoyancy

The buoyancy force in fluids is one of the few forces that is not somehow related to the fluid velocity. It exists in a stationary fluid as well as for one that is moving. It is a force that is directed upwards, or opposite of the weight (which is considered a downward force). A body immersed in a fluid will always experience a buoyancy force. The Greek scientist Archimedes (287 – 212 BC) deduced that the buoyancy force was equal to the weight of the fluid displaced by the body.

The weight of the fluid displaced by a body immersed in it is computed by finding the volume of the object and multiplying it by the density of the fluid (remember, the density is mass per volume) to get the mass of the displaced fluid. This mass is then multiplied by the acceleration due to gravity to get the weight, a force which is then defined as the buoyancy force. If the buoyancy force is greater than the weight of the object, it will float. If the force is less than the weight of the object, it will sink. Because the density of liquids so much larger than the density of gases like air, the buoyancy force for a body immersed in a liquid is much, much higher than that of a body in a gas. Naval architects and ship designers must use the buoyancy forces in their calculations, but the buoyancy forces for airplanes are so small that they are usually ignored. Hot air balloons and blimps do use the buoyancy force to get afloat, but they displace such an extremely large volume of air that the computed buoyancy force exceeds their weight so that they can fly.

11. Streamlines and Flow Patterns

Aerodynamicists and other engineers prefer to visualize the ways fluid moves for a given geometry. They like to know where the flow is going so they can compare experimental and flight test date to the theory. A streamline traces out the path of an element or piece of fluid as it travels in space and time around or through an object. These streamlines are computed mathematically from the velocities in the flow region. Streamlines are usually plotted as smooth lines, and they sometimes have arrows on them to show the direction of the flow.

Streamlines are very useful to illustrate how air moves.

In the section earlier about flow separation, for example, streamlines were used to show how the outside flow traveled around the body and the region of separation and how the flow inside the separation bubble reversed direction. They can be used to show how the air travels around an airfoil (the cross-section or slice of a wing), with some of the air flowing over the top of the airfoil, and the rest flowing below the airfoil.

Vocabulary

object	/ˈɒbdʒɪkt/	n. 物体，实物
wing	/wɪŋ/	n.（鸟、昆虫或蝙蝠的）翅膀，翼；（飞机的）机翼
invisible	/ɪnˈvɪzəbl/	adj. 看不见的，隐形的；无形的，非贸易的；被忽视的
density	/ˈdensəti/	n. 稠密，密集；密度
altitude	/ˈæltɪtjuːd/	n. 海拔高度；高地
airflow	/ˈeəfləʊ/	n. 气流（尤指飞机等产生的）；空气的流动
experimental	/ɪkˌsperɪˈmentl/	adj. 实验性的，试验性的
physics	/ˈfɪzɪks/	n. 物理学；物理特性，物理现象
characteristic	/ˌkærəktəˈrɪstɪk/	n. 特性，特征；特色
supersonic	/ˌsuːpəˈsɒnɪk/	adj. 超音速的 n. 超音速，超声波；超音速飞机
approach	/əˈprəʊtʃ/	v. 靠近，临近；接洽，交谈
friction	/ˈfrɪkʃ(ə)n/	n. 不和，分歧；摩擦；摩擦力

Exercise

Complete the following sentences and translate them into Chinese.

(1) In order to get a good picture of ＿＿＿＿＿＿＿＿＿＿＿＿＿＿＿, for example, the aerodynamicist must know the velocity of the plane, the altitude of the plane, the size and shape of the wing, and the properties of the air.

翻译：＿＿＿＿＿＿＿＿＿＿＿＿＿＿＿＿＿＿＿＿＿＿＿＿＿＿＿＿＿＿＿＿＿＿＿＿

＿＿＿＿＿＿＿＿＿＿＿＿＿＿＿＿＿＿＿＿＿＿＿＿＿＿＿＿＿＿＿＿＿＿＿＿＿＿＿

(2) Sound travels in＿＿＿＿＿＿＿＿＿＿＿＿＿＿＿＿＿＿＿＿（usually air, but sometimes liquid）.

翻译：＿＿＿＿＿＿＿＿＿＿＿＿＿＿＿＿＿＿＿＿＿＿＿＿＿＿＿＿＿＿＿＿＿＿＿＿

＿＿＿＿＿＿＿＿＿＿＿＿＿＿＿＿＿＿＿＿＿＿＿＿＿＿＿＿＿＿＿＿＿＿＿＿＿＿＿

(3) Since both ＿＿＿＿＿＿＿＿＿＿＿＿＿＿＿＿＿＿＿＿＿＿＿＿＿＿，the aerodynamicist must compute the speed of sound at the altitude, temperature, pressure, and density where the plane is flying.

翻译：＿＿＿＿＿＿＿＿＿＿＿＿＿＿＿＿＿＿＿＿＿＿＿＿＿＿＿＿＿＿＿＿＿＿＿＿

＿＿＿＿＿＿＿＿＿＿＿＿＿＿＿＿＿＿＿＿＿＿＿＿＿＿＿＿＿＿＿＿＿＿＿＿＿＿＿

(4) The Mach number is named for Ernst Mach (1838 – 1916), who conducted the first _____ while a professor of experimental physics at the University of Prague.

翻译：_____
_____

(5) As the plane _____ (Mach number greater than 1.0), the waves compress into a cone-shaped envelope about the plane.

翻译：_____
_____

(6) Anything that _____ must overcome the resistance to motion between the two objects.

翻译：_____
_____

(7) In general, the boundary layer gets _____.

翻译：_____
_____

(8) The flow inside the boundary layer will be _____, meaning that the layers will basically stay in layers, without mixing.

翻译：_____
_____

╎ **Supplementary Reading** ╎

Bernoulli's Theorem

In order to better understand the changing flow field around an object, aerodynamicists needed a formula for the relationship between the velocities in the fluid and the pressures. Daniel Bernoulli (1700 – 1782) was the first to calculate an expression relating the terms (Figure 2.2). For steady (unchanging in time), inviscid (no friction forces), incompressible (density is constant) flow, he found that the local pressure plus one half times the density times the local velocity squared was constant along a streamline. This meant that a particle on a streamline would have the same sum all along the streamline.

In front of an airfoil, for example, the particle would be at the freestream pressure, density and velocity. The sum of the pressure plus one half times the density times the velocity squared could be computed. As the particle traveled up over the airfoil surface, its velocity would change, it would speed up as it turned up over the front of the airfoil, then

Figure 2. 2   Daniel Bernoulli (1700 – 1782)

slow down as it traveled down the back side of the airfoil. As it moved away from the airfoil trailing edge, it would have returned to the freestream velocity value. The Bernoulli equation shows that as the velocity changed, the pressure would have to change, as well, since the sum computed from the freestream values must remain the same.

Over time, other famous mathematicians and aerodynamicists have expanded the relationship between the pressures and the velocities in a fluid flow, but all show the same the basic behavior as Bernoulli's Theorem; when the velocity in the flow increases, the pressure decreases, and when the velocity decreases, the pressure in creases.

As discussed in the Math number section, when a plane flies faster than the speed of sound, the waves of changing pressure are compressed in to a single, infinitesimally thin layer. This is called a shock wave, and fluid properties such as pressure, density, temperature, and velocity change drastically and instantaneously through a shock wave.

Theoretically, once formed a shock wave travels on to infinity. In nature, however, atmospheric winds cause the shock to weaken and disperse. When an aircraft flying at supersonic speeds is at a high altitude, the shock wave is diffused long before it travels to the earth's surface. If a supersonic airplane flies too close to the ground, however, the shock will hit the earth's surface. It will be heard by observers on the ground (it's called a sonic boom), and if the shock is strong enough, it will cause building to shake and windows to break.

The space shuttle has a shock wave around it as it returns to earth through the atmosphere. There is a section of southwestern Georgia that is along the flight path of the returning shuttle when it lands at Cape Canaveral.

When the shuttle travels along this path, it is still slightly supersonic, and it is close enough that the people on the ground hear the sonic boom as it travels overhead. The shuttle can't be seen, but it can be heard! Before the shuttle flies low enough for the shock to cause any damage, however, it has dropped its speed below Math 1.0 and the shock is gone.

In the early days of flight, the aerodynamics of transonic and supersonic flight were not well understood. When pilots started flying faster and faster, their planes were not designed for the rigors of sonic flight. The fronts of the planes and the wings were relatively fat and blunt (rounded), which was fine for subsonic flight. As these early planes approached the sonic regime (called the sound barrier in the those days), the pressure waves compressed together to form shock waves in the supersonic regions over the plane, causing vibrations and buffeting (bumpiness). Sometimes the vibrations and buffeting were enough to cause the planes to break up in flight! People were convinced that there was an invisible solid barrier at Mach 1.0, and that humans were not intended to go faster than the speed of sound.

In the late 1940's, designers started to understand high speed aerodynamics and began to design aircraft to fly in the supersonic regime. They designed planes that had sharp and thin airfoil shapes for the wings. This allowed the designers to control the strength of the shock waves that formed to sharply decrease the vibrations and buffeting. On October 14th, 1947, flying the experimental aircraft Bell XS-1, Captain Charles Yeager flew the first successful supersonic flight. Today, many pilots regularly fly faster than the speed of sound in planes designed for transonic and supersonic flight. The transition to supersonic is so smooth they only notice it because of changes in their instruments.

Lesson 3   Forces of Flight

The flight of an airplane, a bird, or any other object involves four forces that may be measured and compared: lift, drag, thrust and weight. As can be seen in the Figure 3.1 below, in straight and level flight these four forces are distributed with the lift force pointing upward, opposite to the weight, the thrust pointing forward in the direction of flight, and the drag force opposing the thrust. In order to fly, the lift force must be greater than or equal to the weight, and the thrust force must be greater than or equal to the drag force.

Figure 3.1   Direction of Forces in Straight and Level Flight

1. Weight

The weight of the aircraft is equal to its mass times the acceleration of gravity. It is a natural force, and it is a measure of the force that pulls the plane down towards the earth. Therefore, the direction assigned to the weight is downward (Figure 3.1).

2. Lift

The force that pushes an object up against the weight is lift. On an airplane or a bird, the lift is created by the movement of the air around the wings.

If two particles were released from the same point at the same time, one on each streamline, they would start out moving together. As they approached the front of the airfoil, however, their velocity will start to change. Initially, each will start to move faster; their velocities will increase. As they turn back downward along the back half of the airfoil, they will slow back down to their initial freestream velocity. Due to the shape of the airfoil, the air moves faster over the top of the airfoil then it does on the lower surface. The faster air leads to a lower pressure (form Bernoulli's low) on the upper surface and hence a net force is produced.

When the total force on each surface is computed, the difference in the pressures between the upper and lower surfaces will create a smaller force on the upper surface, and a larger force on the lower surface. The smaller force will be pointed downward, and the larger force will be pointing upwards. When the two forces are combined, the net force (lower force minus the upper force because the directions are opposing) is the lift force, and it will be directed upwards.

The shape of the airfoil (wing) is important for lift, and is designed carefully. Most airfoils today have camber, or curved upper surfaces and flatter lower surfaces. These airfoils generate lift even when the flow is horizontal (flat). The Wright brothers used symmetric airfoils in their airplane design. Since the upper and lower surfaces were the same the pressure on the either surface (top or bottom) are the same, so the net combined force on the airfoil is aero and there is no lift! How, then, did the Wright brothers get their airplane off the ground?

In order to generate lift with a symmetric airfoil, the airfoil must be turned (tilted) with respect to the flow, so that the upper surface is "lengthened" and the lower surface is "shortened". This turning with respect to the flow is called the angle of attack. A pilot can increase the lift for both cambered and symmetric wings by increasing the angle of attack of the wing with respect to the freestream flow. This is why an airplane rotates slightly at take-off; the pilot is increasing the angle of attack to generate more lift. If the angle of attack is doubled, the lift doubles. There is a limit to how much lift can be generated, however. The angle of attack can be increased to a point where the flow on the upper surface separates. The streamline above the upper surface now travels over the separation bubble. The pressures on the upper surface suddenly increase, and the net lift force drops drastically.

Airflow deflection is another way to explain. To understand the deflection of air by an airfoil let's apply Newton's Third Law of Motion. The airfoil deflects the air going over the upper surface downward as it leaves the trailing edge of the wing. According to Newton's Third Law, for every action there is an equal, but opposite reaction. Therefore, if the airfoil deflects the air down, the resulting opposite reaction is an upward push. Deflection is an important source of lift. Planes with flat wings, rather than cambered, or curved wings must tilt their wings to get deflection.

Another way to increase the lift on a wing is to extend the flaps. This again lengthens the upper surface and shortens the lower surface to generate more lift.

The velocity of the freestream air (actually of the airplane) is the most important element in producing lift. If the velocity of the airplane is increased, the lift will increase dramatically. If the velocity is doubled, the lift will be four times as large.

The generation of lift is also used in other applications. Race car designers use airfoil-like surfaces to generate negative lift, or a downward-directed force. This force combined with the weight of the race car helps the driver maintain stability in the high-speed turns on the race track. In other words, the additional downward facing force helps offset the opposing forces in the turns.

Lift on an airplane or a bird is primarily generated by the wings. The amount of lift generated by the body, or fuselage, of the plane or bird and the tail is very small. So, the lift force generated by the wings is the basic lift force on the airplane. Its direction is upward, opposite the weight. If the lift is greater than the weight, the plane will fly. After the plane has climbed to the cruising altitude, it will level off (decrease the angle of attack to zero), and the lift force will be equal to or very slightly larger than the weight. If the lift force is smaller than the weight, the plane will lose altitude and return to earth.

3. Thrust

A forward direction force called the thrust is generated by the engines of the airplane (or by the flapping of a bird's wings). The engines push high velocity air out behind the plane, and the difference between the high velocity exhaust gases and the original velocity of the airplane creates the forward directed thrust. Because its direction is perpendicular to the forces of weight and lift, the thrust force is unaffected by either.

4. Drag

The drag is the fourth of the major forces for flight. It is a resistance force to the forward motion of any object, including planes. There are four types of drag: friction drag, form drag, induced drag and wave drag and they are functions of the shape of the body, the smoothness of the surfaces and the velocity of the plane. All four sum together for the overall drag force. Since the drag force resists the motion of the plane its direction is opposite the plane goes forward, but if the drag force exceeds the thrust, the plane will slow down and stop.

The friction drag is sometimes also called the skin friction drag. It is the friction force at the surfaces of the plane caused by movement of air over the whole plane. In the boundary layer along each surface, a skin friction coefficient can be computed as a function of the velocity of the air and the surface roughness. The summation of all the local skin friction coefficient is used to compute the friction drag. Aerodynamicists design the outside of the airplanes to be smooth surfaces so that the friction coefficients and therefore the friction drag will be small. If a person were to look at the surface of a wing, for example, he or she would see that all the sheets of metal join smoothly, and even the rivets are rounded over and are as flush with the surface as possible. Sometimes the aerodynamicist will design small tabs to be placed along a wing surface to trip the boundary layer for transition to turbulence. While this increase the friction drag slightly, the increase in lift and control of the airplane are judged worth the cost.

The form drag, or pressure drag, as it is sometimes called, is directly related to be the shape of the body or airplane. A smooth, streamlined shape will generate less form drag than a blunted or flat body. The term streamlined comes from the idea that a shape is designed so that streamlines above and below the body barely change and rejoin smoothly right behind the body. A thin, relatively sharp-nosed airfoil is a perfect example of streamlining.

Any object that moves through a fluid can get a decrease in form drag by streamlining. Automobiles are streamlined, which translates into better gas mileage; there is less drag, so less fuel is required to "push" the car forward. Busses, vans and large trucks are less streamlined, and this is one reason (the additional weight is another) why they use more fuel than smaller, streamlined cars.

Form drag is easy to demonstrate using a hand out the window of a moving car. If the hand is held flat and horizontal to the ground, it is basically a streamlined object, and the observer feels only a small tug, or drag. If he or she turns the hand so that the palm is facing forward, the drag force is greatly increased, and the hand is pulled backwards! If the observer could see what was happening in the invisible air, he or she would see many little swirls of air, or eddies, behind the hand. The streamlines around the hand would have to travel around these eddies, the pressures would be higher along the streamlines, and the drag is increased dramatically.

An interesting offshoot of the discussion of form drag is in the design of large banners like those used to advertise school fairs, church picnics, or museum openings. If a large cloth or plastic banner is strung between two buildings across a street, and the wind blows against it, it has a high form drag. This form drag can be so large that the banner rips and tears, and it may be destroyed. Banner designers now know that they must include slits in the fabric to allow the air to move through the banner, so that the drag on the banner is decreased.

The first two types of drag are often added together and called the profile drag by aerodynamicists. Pilots call the sum of these two drags the parasite drag. In either case， these drags are primarily a function of the shape of the body and the smoothness of the surfaces. All objects moving through a fluid will have these drag forces. Airplanes， because of the amount of lift generated and the velocities at which they travel， are also subject to two additional drag forces， the induced drag and the wave drag.

Induced drag is sometimes called the drag due to lift. As the lift force is generated along a wing， a small amount of excess force can be generated in the direction opposing the motion， or in the direction of drag. This excess is called the induced drag， and because of its direction， it causes a decrease in the plane's motion. Therefore， it is considered a type of drag force. It is one of the odder concepts that an aerodynamicist must consider during the design of an airplane. Any change that he or she can make to increase the lift is a positive change. In the design of a better wing a designer will optimize the generation of the lift so as to minimize the generation of induced drag； a well-designed wind will generate the needed lift， while minimizing the induced drag. However， other factors such as structural strength， overall weight， cost and complexity will also control the design of the "optimal" wing.

The last of the four types of drag is the wave drag. This generally only occurs when the airplane is flying faster than the speed of sound in supersonic flight. It is caused by the interactions of the shock waves over the surfaces and the pressure losses due to the shocks. Wave drag can also occur at transonic speeds， where the velocity of the air is supersonic locally. Since most commercial jests today fly at transonic speeds， wave drag is an important part of total drag.

Vocabulary

force	/fɔːs/	n. 暴力，武力；力，力量；自然力
measure	/ˈmeʒə(r)/	v. 测量
straight	/streɪt/	adj. 直的，笔直的
distributed	/dɪˈstrɪbjuːtɪd；	adj. 分布式的，分散式的
	ˈdɪstrɪbjuːtɪd/	
thrust	/θrʌst/	n. 要点，要旨；猛推，刺；（飞机、火箭等的）驱动力，推力

acceleration	/əkˌseləˈreɪʃ(ə)n/	n. 加速，加快；(车辆)加速能力；加速度
gravity	/ˈɡrævəti/	n. 重力，地心引力
lift	/lɪft/	n. 电梯，升降机；(空气的)升力，提升力
airfoil	/ˈeəfɔɪl/	n. 机翼，翼型
particle	/ˈpɑːtɪk(ə)l/	n. 微粒；极少量；粒子；(数)质点
velocity	/vəˈlɒsəti/	n. 速度，速率
drag	/dræɡ/	n. 阻力，抗力

Exercise

Complete the following sentences and translate them into Chinese.

(1) The flight of an airplane, a bird, or any other object involves four forces that may be measured and compared; _____.

翻译：_____

_____

(2) The weight of the aircraft is _____.

翻译：_____

_____

(3) The force that _____ is lift.

翻译：_____

_____

(4) Due to the shape of the airfoil, _____ then it does on the lower surface.

翻译：_____

_____

(5) When the total force on each surface is computed, the difference in the pressures between the upper and lower surfaces will _____, and a larger force on the lower surface.

翻译：_____

_____

(6) In order to _____, the airfoil must be turned (tilted) with respect to the flow, so that the upper surface is "lengthened" and the lower surface is

"shortened".

翻译：_____

_____

(7) A forward direction force called the thrust is _____ of the airplane (or by the flapping of a bird's wings).

翻译：_____

_____

(8) The drag is _____ the major forces for flight.

翻译：_____

_____

(9) The friction drag is sometimes also called _____.

翻译：_____

_____

(10) Any object that _____ can get a decrease in form drag by streamlining.

翻译：_____

_____

✛ Supplementary Reading ✛

　　Every pilot has a working knowledge of these four basic forces of flight, and he or she uses this knowledge to fly the plane. Aerobatic pilots, in particular, are constantly balancing these forces, and using the concepts in this chapter to design amazing stunts to delight the crowds. They will deliberately stall the wings of the airplane to cause the plane to lose lift and drop suddenly. They very carefully fly upside down, balancing the new lift force with the weight of the plane. They will point the airplane straight up into the air and fly straight up as far as they can, let the plane hang there for a second, then let if fall back down its original path. In this configuration, the lift force now points to the right, the thrust force points straight up, and the weight and drag forces point downward towards the earth. The drag and weight and drag forces point downward towards the earth. The drag and weight forces together exceed the thrust, and the higher the plane flies, the less thrust is generated. Eventually, they are even, and the plane seems to hang there for a second or two. Then the weight and drag forces dominate, and the plane drops backwards. After a few heartbreaking seconds, the pilot will turn the airplane back so the nose points downward into the direction of the flow, increase the thrust, and re-establish the original force balances for level flight.

 *Section 2*

Aircraft

Lesson 4   Parts of an Aircraft

One of the main parts of the aircraft is the fuselage. On C919, it consists of the cockpit, the passenger cabin and the cargo compartments. On cargo planes, the cabin is replaced by a cargo compartment. Fuel is contained in three tanks in the wings and wing center section. In some versions of the planes, an auxiliary tank can be installed in a part of the cargo compartment.

The fuselage is formed from circumferential frames, longitudinal stringers and skin. The fuselage is divided horizontally by the floor, which is built up from beams and panels.

The wings are another main part of the aircraft. Basic wing structure consists of left, center and right wing sections. These are built up from the front spar, rear spar, ribs, top and bottom skins, and stringers. The wing ribs in the tanks prevent a surge effect in the fuel tank.

There are various control surfaces located on the wings. On most aircraft, leading edge flaps and slats are located on the front of the wings. Leading edge flaps are sometimes called Krueger flaps. There are also inboard and outboard flaps on the trailing edge. These control surfaces are hydraulically powered and are only extended or retracted during take-off and landing.

On a modern aircraft, there are inboard ailerons for roll control and outboard ailerons for low airspeed flight. The upper surface of the wing also contains spoilers. They are operated by hydraulic pressure and are used for braking.

Another main part of the aircraft is the tail section. This consists of the vertical stabilizer, sometimes known as the fin, and the horizontal stabilizer. Longitudinal trim is accomplished by movement of the horizontal stabilizer. The elevators are attached to the trailing edge of this stabilizer and provide additional pitch control. The rudder is used for directional and yaw control.

The last two main parts of the plane are the engines and the landing gear. As a rule, the engines are attached under the wings, but, in the case of the modern aircraft, the two side engines are mounted on struts outboard of the fuselage. The center engine is mounted aft of the fuselage structure below the vertical fin. The inlet for the center engine is above the fuselage, and air intake is through an S-shaped duct.

The landing gear consists of the main gear which is attached to the wing rear spar and retracts to stow under the main cabin aft of the wing rear spar. The nose gear is retracted forward into a wheel well in the lower nose compartment. The nose gear is used for maneuvers on the ground. The landing gear is retracted and extended hydraulically.

Vocabulary

aircraft	/ˈeəkrɑːft/	n. 飞机，航空器
fuselage	/ˈfjuːzəlɑːʒ/	n. 机身
cockpit	/ˈkɒkpɪt/	n. 驾驶舱
compartment	/kəmˈpɑːtm(ə)nt/	n. 舱位
cabin	/ˈkæbɪn/	n. 客舱
cargo	/ˈkɑːgəʊ/	n. 货舱
auxiliary	/ɔːgˈzɪlɪərɪ/	adj. 辅助的
circumferential	/səˌkʌmfəˈrenʃəl/	adj. 周向的
longitudinal	/ˌlɒn(d)ʒɪˈtjuːdɪn(ə)l/	adj. 纵向的
stringer	/ˈstrɪŋə/	n. 长桁
leading edge flap		前缘襟翼
slat	/slæt/	n. 缝翼
Krueger flaps		克鲁格襟翼
trailing edge flaps		后缘襟翼
extend	/ɪkˈstend/	v. 放出
retract	/rɪˈtrækt/	v. 收回
aileron	/ˈeɪlərɒn/	n. 副翼
spoiler	/ˈspɒɪlə/	n. 扰流板
tail section		尾部
rear	/rɪə/	adj. 后部的
vertical stabilizer		垂直安定面
horizontal stabilizer		水平安定面

Exercise

1. Make sentences with the words given below.

Example：

The C919—manufacture—the China

The C919 is manufactured in the China.

(1) Airbuses—manufacture—Europe

(2) The planes—check—after each flight

(3) Maintenance personnel—train—Air China Training centre

(4) This aviation school—build—1970

(5) The 747—overhaul—in Ameco last May

2. Translate the following sentences into Chinese.

(1) Scheduled services were introduced in 1955.

翻译：

(2) The 747 was overhauled in Beijing last year.

翻译：

(3) The A-check was included in this check.

翻译：

(4) All major components were replaced during the overhaul.

翻译：

(5) The plane was repainted three weeks ago.

翻译：

(6) Some modifications were carried out yesterday.

翻译：

(7) That plane was overhauled in Ameco.

翻译：

(8) The school was opened 21 years ago.

翻译：

(9) The first Airbus was delivered in February 1976.

翻译：

(10) The tire pressure was checked before the flight.

翻译：

(11) The airport building was extended last year.

翻译：

(12) A new 737−800 was delivered last year.

翻译：

The Boeing 747

The Boeing 747 (Figure 4.1) marked the beginning of a new era in air transport, the Jumbo-age. The Jumbo carries about three times as many passengers as a Boeing 707 and its cargo capacity is three times the capacity of the 707's. The first Air China Jumbo went into service in April 1978 and there are about ten of these wide-body jets with the flying crane symbol on their fins. Air China has replaced the older 747's with the latest models which need less fuel and have increased engine power.

Figure 4.1　Boeing 747

The four huge engines consume about 14, 500 liters of kerosene per hour. The air intake of the engines has a diameter of 2.19 meters. To mention a couple of other facts, the artificial horizon shows the pilot the aircraft's position in flight at any time. This is particularly important when visibility is low due to clouds or haze. The aircraft's weather radar equipment tells the pilot whether the clouds ahead are harmless or whether they hide possible turbulence.

There are three different versions of the 747. They all have a wing-span of 59.64 meters and a length of 70.51 meters. The take-off weights, however, vary from 333 to 362 metric tons. The Air China passenger versions have room for up to 361 people to be carried over a maximum distance of 9, 800 kilometers. The combi-version, the 747 SL, has 249 seats and an additional freight compartment for six freight pallets aft of the passenger cabin. A special side cargo door allows quick and easy access to this cargo compartment. All Jumbos provide further stowage for freight in their bellies underneath the cabin.

The giant air freighter, which is loaded from the front by means of a hinged nose which moves upwards, and a side cargo door in the rear, was operated by Air China in 1983. This Jumbo freighter can carry about 100 metric tons of freight. The weight, which can be up to 362 metric tons, is supported by four main landing gears with four wheels each and a nose gear with two wheels. This means that each set of wheels puts less weight on the runway than those of a 707, which has half the total weight of the Jumbo.

Lesson 5　Electric Power System

1. Generation

Each of the 4 engines drives an Integrated Drive Generator (IDG) providing alternating current power. The APU (Auxiliary Power Unit) drives a fifth generator (APU GEN) which can supply all the aircraft network on ground or replace a main generator in flight.

Two Transformer Rectifiers (TR) utilize the alternating current to supply the main direct current network. Another transformer rectifier (ESS TR) supplies the essential direct current network. Two batteries also supply direct current in some configurations. The APU TR or/and a battery (APU BAT) power supply the APU starter, depending on the aircraft configuration. In case of major failure, a constant speed hydraulic motor drives an emergency generator: both equipment forms an assembly called Constant Speed Motor/Generator (CSM/G). The function of the CSM/G is to supply the systems necessary to the aircraft control.

2. General Arrangement of the Distribution Network

1) AC Distribution Network

There are four distribution networks:

• Networks 1 and 2 associated to left side (side 1) generators (respectively IDG 1 and 2);

• Networks 3 and 4 associated to right side (side 2) generators (respectively IDG 3 and 4).

In normal flight configuration each IDG supplies its own distribution network via its line contactor (GLC).

The distribution network 1 consists of:

• The alternating current bus 1XP1 (AC BUS 1 - 1);

• The alternating current essential bus 9XP (AC ESS BUS);

• The alternating current sheddable essential bus 4XP (AC SHED ESS BUS).

The distribution network 2 consists of:

• The alternating current bus 1XP2 (AC BUS 1 - 2).

The distribution network 3 consists of:

• The alternating current bus 2XP3 (AC BUS 2 - 3).

The distribution network 4 consists of:

• The alternating current bus 2XP4 (AC BUS 2 – 4).

2) DC Distribution Network

The TR1 powered from the AC BUS 1 – 2 supplies through its contactor:

• The direct current bus 1PP (DC BUS 1);

• The direct current bus 3PP (DC BAT BUS).

The TR2 powered from the AC BUS 2 – 3 supplies through its contactor:

• The direct current bus 2PP (DC BUS 2);

• The direct current service bus 6PP (DC SERVICE BUS).

The ESS TR powered from AC BUS 1 – 1 supplies through its contactor:

• The direct current essential bus 4PP (DC ESS BUS);

• The direct current sheddable essential bus 8PP (DC SHED ESS BUS).

Two batteries are associated with the DC BAT BUS through their own contactor. They can also supply the static inverter during particular configurations.

3) Particular Configurations

The AC ESS BUS and the AC SHED ESS BUS plus, via the ESS TR, the DC ESS BUS and the DC SHED ESS BUS can be supplied by:

• The AC BUS 2 – 4 in case of loss of the AC BUS 1 – 1;

• The CSM/G more especially in emergency configuration.

Vocabulary

Integrated Drive Generator (IDG)		整体传动发电机
Auxiliary Power Unit (APU)		辅助动力系统
transformer rectifier		变压整流器
static inverter		静变流机
independent	/ˌɪndɪˈpendənt/	*adj.* 独立的
alternating current		交流电
direct current		直流电
Constant Speed Motor/Generator		恒速液力马达/发电机
bus	/bʌs/	*n.* 汇流条
battery	/ˈbæt(ə)rɪ/	*n.* 电瓶、蓄电池

Exercise

Fill in the blanks with the correct words and, if necessary, change the grammatical form.

consist of   either   enable   failure   linked

provide   disconnect   above   indicate   pull

(1) If there is a _____ in one of the systems, the plane must be grounded.

(2) The cockpit crew _____ a captain, a first officer and—on some planes—a flight engineer.

(3) Before you start the test, make sure you _____ electrical power.

(4) A knowledge of the English language _____ you to understand the Manuals.

(5) The pressure must not be _____ 25 psi.

(6) Before you _____ the IDG system, read the instructions carefully.

(7) The opposite of "push" is _____.

(8) The computer _____ the malfunctions of the system.

(9) With the help of a contactor the busses can be _____.

(10) "You can choose _____ procedure" means that is doesn't matter which one you choose.

Supplementary Reading

Electrical Power

1. Purpose

The electrical power system makes and supplies AC and DC power to airplane. The system has automatic and manual controls and protection. A standby AC and DC system gives normal and emergency power.

2. AC Power

The electrical power system has four main AC power sources and one standby power source. These are the main AC power sources and their supply capacity:

- Left integrated drive generator (IDG 1) (90 KVA);
- Right integrated drive generator (IDG 2) (90 KVA);
- APU starter-generator (90KVA below 32,000 feet/9,753 meters, and goes down to 66 KVA at 41,000 feet/12,496 meters);
- External power (90 KVA).

The IDGs and APU starter-generator supply a 3 phase, 115/200 volts (nominal) at 400 Hz. The AC power system design prevents two sources to the same load at the same time. The static inverter supplies a one phase, 115 V AC output to the AC standby bus.

3. DC Power

Three transformer rectifier units (TRUs) change115 V AC to 28 V DC. The airplane also has these DC power sources:

4. Battery

The battery is the backup DC source if other sources do not operate. The standby power control unit (SPCU) controls the distribution of DC power.

5. Standby Power

With the loss of normalpower, the standby power system supplies a minimum of 30 minutes of AC and DC power to systems necessary to maintain safe flight. The battery supplies DC power. The static inverter uses battery power to make AC power. The SPCU controls the distribution of AC and DC standby power.

Lesson 6　Hydraulic System

1. General

A. When checking for leakage, pressurize applicable systems and examine all components, hydraulic fittings and connecting lines. Pin hole leaks may be detected by noting damp spots on the airplane and tracing run back fluid to locate source of leak. When disconnecting hydraulic lines during removal of components take care to prevent spillage of hydraulic fluid. Should fluid leak on the airplane or spill while performing maintenance, the affected area must be decontaminated (Ref 12 – 40 – 00).

B. Leakage in the standby hydraulic system will be reflected in the loss of hydraulic fluid from System B hydraulic reservoir.

C. Leakage at the Power Transfer Unit pump and associated plumbing will be reflected in the loss of hydraulic fluid from system B reservoir. Leakage of the Power Transfer Unit motor and associated plumbing will be reflected in the loss of fluid from System A.

D. Illumination of any pump low pressure warning light during operation of a system and a confirmation of loss of pressure on the affected pump should be followed by a check of the corresponding pump pressure filter and case drain return filter. Contaminated filter elements indicate a defective pump. A pump need not be operated again following warning light illumination unless filter check shows no sign of contamination and trouble is suspected outside the pump. If additional information is required on the pump low pressure warning lights, refer to 29 – 34 – 01.

E. Transfer of hydraulic fluid between hydraulic systems A and B is caused by improper operation of system controls or worn equipment. See trouble shooting steps below for correct operation or replacement of parts.

F. When trouble shooting, it is not necessary to perform a check of both systems. Only the system (A or B) where the trouble is present needs to be checked. By feeling for hot tubing or actuators or listening for fluid leakage, faulty components can be isolated within a subsystem which has excessive internal leakage. Whenever possible, standard tools for detecting heat, vibration or sound should be used. Before checking for internal leakage by these methods, cycle the components to be checked to ensure that personnel will not be injured or equipment damaged when the component moves.

G. When trouble shooting, the hydraulic system should stabilize between 2, 800 and

3,200 psi with flight control and spoiler switches ON and 2,900 to 3,200 with flight control and spoiler switches OFF.

H. If either engine No.1 or 2 fire switches are positioned to FIRE with the corresponding engine-driven pump operating, proceed with the following steps:

(1) If a supply shutoff valve was closed for less than 5 minutes, no maintenance is required and shutoff periods are not accumulative.

(2) If supply shutoff valve was closed for longer than 5 minutes, ensure that pump has operated at least one minute with supply fluid available before checking filter elements so that metal contamination, possibly generated during fluid shutoff, has reached the filter.

① Check case drain filter element for metal contamination (Ref 29 – 15 – 91, MP).

② Check pressure filter for metal contamination (AMM 29 – 15 – 71/201).

③ If no metal is found in either filter, install new filters and continue to run pump. Recheck filters in 200 hrs.

④ If metal is found, remove applicable pump for overhaul and flush lines between the pump and filters (Ref I/C).

2. Prepare for Trouble Shooting

A. Service hydraulic reservoir(s) (AMM 12 – 12 – 00/301).

　　NOTE: Standby system is serviced along with System B.

B. Pressurize hydraulic reservoir(s) (AMM 29 – 09 – 00/201).

C. Provide electrical power (AMM 24 – 22 – 00/201).

D. Check that parking brake is set.

E. When trouble shooting engine-driven pumps, pressurize system by operating or motoring engine (AMM 71 – 00 – 00/201).

F. When trouble shooting using an electric pump, position applicable ELEC hydraulic pump switch (P5) ON.

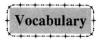
Vocabulary

pressurize	/ˈpreʃəraɪz/	v. 增压
hydraulic fittings		液压配件
confirmation	/kɒnfəˈmeɪʃ(ə)n/	n. 证实
detect	/dɪˈtekt/	v. 探测
Power Transfer Unit (PTU)		动力转换单元
pump	/pʌmp/	n. 泵

pressure filter		压力过滤器
case drain return filter		壳体回油过滤器
flush	/flʌʃ/	v. 冲洗
illumination	/ɪˌljuːmɪˈneɪʃən/	n. 照明
improper	/ɪmˈprɒpə/	adj. 不正确的，不适合的
isolate	/ˈaɪsəleɪt/	v. 隔离
filter element		滤芯
vibration	/vaɪˈbreɪʃ(ə)n/	n. 震动
Engine-driven Pump		发动机驱动泵
supply shutoff valve		供应关断阀门
hydraulic	/haiˈdrolik/	adj. 液压的
secondary	/ˈsek(ə)nd(ə)rɪ/	adj. 辅助的，次要的
thrust	/θrʌst/	n. 推力
reservoir	/ˈrezəvwɑ(r)/	n.（液压）油箱
driven	/ˈdrɪvn/	adj. 驱动的
filter	/ˈfɪltə/	n. 过滤器；油滤

Exercise

1. Fill in the blanks with the following words.

additional　　confirmation　　examine　　excessive　　faulty

improper　　loss　　performed　　pressurized　　worn

(1) Fill up the system to the yellow line if there has been a _____ of fluid.

(2) When checking the surface _____ the rivets for signs of corrosion.

(3) This job must not be _____ without using special tools.

(4) The tank must be _____ prior to performing a leak test.

(5) The damage was due to _____ use of the pump.

(6) The part can withstand high temperatures but _____ heat will certainly damage it.

(7) _____ information can be found in Chapter 17.

(8) Any _____ of faulty wiring must be dealt with immediately.

(9) The _____ pumps were replaced and a subsequent check was performed.

(10) The system was checked and the part was found to be _____.

2. Answer the following questions according to the passage.

(1) What is the main function of the hydraulic system?

(2) What is the filter used?

(3) How many sub-systems are included in the hydraulic power system?

(4) Which hydraulic system can supply hydraulic power to leading edge flaps and slats?

(5) How do you operate the hydraulic system?

3. Translate the following sentences into Chinese.

(1) The hydraulic reservoir is a tank or a container to stone sufficient hydraulic fluid for all conditions of operation.

翻译：_____

(2) Hydraulic power is transmitted by movement of fluid by a pump. The pump does not create the pressure, but the pressure is produced when the flow of fluid is restricted.

翻译：_____

(3) Air pressure from the reservoir pressurization system maintains head pressure on hydraulic system A, B, and the standby hydraulic system reservoirs. The pressurized reservoirs supply a constant flow of fluid to the hydraulic pumps.

翻译：_____

(4) Before you operate the pump again, let the reservoir temperature decrease to ambient temperature or damage to equipment can occur.

翻译：_____

(5) Normally, the Engine-Driven Pumps are on. When the engines are on, the Engine-Driven Pumps come on to also pressurize systems A and B.

翻译：_____

┊ **Supplementary Reading** ┊

Hydraulic Power

1. General

There are three independent hydraulic systems that supply hydraulic power for user systems.

The main and auxiliary hydraulic systems supply pressurized fluid to these airplane systems:

- Both thrust reversers;
- Power transfer unit (PTU) motor;
- Landing gear extension and retraction;
- Nose wheel steering;
- Main gear brakes;
- Primary flight controls;

• Secondary flight controls.

These systems make up the hydraulic power system：

• Main hydraulic systems；

• Ground servicing system；

• Auxiliary hydraulic systems；

• Hydraulic indicating systems.

2. Main Hydraulic Systems

The main hydraulic systems are A and B. System A has most of its components on the left side of the airplane and system B on the right side.

3. Ground Servicing System

The ground servicing system fills all hydraulic reservoirs from one central location.

4. Auxiliary Hydraulic Systems

The auxiliary hydraulic systems are the standby hydraulic system and the power transfer unit (PTU) system.

The standby hydraulic system is a demand system that supplies reserve hydraulic power to these components：

• Rudder；

• Leading edge flaps and slats；

• Both thrust reversers.

The hydraulic power transfer unit (PTU) system is an alternate source of hydraulic power for the leading-edge flaps and slats and auto slat system.

5. Hydraulic Indicating Systems

These are the indicating systems：

• Hydraulic fluid quantity；

• Hydraulic pressure；

• Hydraulic pump low pressure warning；

• Hydraulic fluid overheat warning.

Lesson 7   Pneumatic System

A. The purpose of the pneumatic system is to supply bleed air from the 5th and 9th stage of the engine compressor or from the Auxiliary Power Units (APU) or from a ground cart. Air is distributed by a pneumatic manifold from the above sources to the air conditioning packs, wing and cowl Thermal Anti-Ice (TAI) systems, the engine starting system, potable water system and the hydraulic reservoir. The systems using air from the pneumatic manifold are covered in their respective chapters.

B. The main supply of bleed air to the manifold is obtained from the engine 5th stage compressor. 5th stage air is supplied to the pneumatic manifold at all times except at engine idle or at low engine thrust settings. When this occurs 9th stage bleed air is automatically used. Air from the 5th and 9th stage bleed ports is controlled by the engine bleed control system described in Chapter $36-11-05$. APU bleed air or pneumatic ground cart bleed air is primarily used for engine starting and for air conditioning pack operation on the ground. The APU can be used as an alternate bleed air source up to airplane altitudes of 17, 000 ft.

C. The pneumatic manifold interconnects the engine bleed system and APU and distributes bleed air to the systems outlined in paragraph A. The pneumatic manifold contains the necessary valves to shut off bleed air at each engine, APU and to isolate the left and right hand systems.

D. Each engine bleed system contains a bleed air temperature control sub-system. This consists of a precooler, precooler control valve and precooler control valve sensor. The engine bleed air temperature is automatically controlled to a preset temperature whenever the engine is operating and the Pressure Regulator and Shutoff Valve (PRSOV) is open. The temperature is sensed in the precooler bleed air discharge. The sensor will cause the precooler control valve to modulate open/closed as required to maintain this temperature. Engine fan air is used as the heat sink for the bleed air.

E. Two pressure transmitters are provided for pressure indication of the bleed air. One transmitter is used for each engine. Both pressure transmitters are connected to a dual pressure indicator on the overhead panel. The engine bleed air overtemperature indication consists of overtemperature switches in the ducting system, which are connected to the overhead panel trip lights. The temperature switches illuminate the trip lights and the corresponding engine PRSOV closes when the bleed air temperature exceeds approximately 490 °F. Also, the bleed trip light will illuminate if there is excessive pressure at the bleed valve inlet.

F. The high-stage manifold duct on the engine has a bleed air pressure tap provision to supply pneumatic pressure to airplanes equipped with potable water engine pressurization system.

(1) AIRPLANES WITH POTABLE WATER ENGINE PRESSURIZATION SYSTEM.

Make sure that the potable water pressure line is connected to the pressure tap on engine 1 (AMM 71 – 00 – 02/401).

(2) AIRPLANES WITHOUT POTABLE WATER ENGINE PRESSURIZATION SYSTEM.

Make sure that the pressure tap provision on the replacement engines are sealed (AMM 71 – 00 – 02/401).

Vocabulary

pneumatic	/njuː'mætɪk/	n. 气动的
engine compressor		发动机压气机
thermal	/'θɜːm(ə)l/	adj. 热的
anti	/'æntɪ/	adj. 防……的,抗……的
bleed	/bliːd/	n. 引气
duct	/dʌkt/	n. 管道
available	/ə'veɪləb(ə)l/	adj. 可用的,有效的
adequate	/'ædɪkwət/	adj. 足够的,充足的
isolation	/aɪsə'leɪʃ(ə)n/	n. 隔离
precooler	/priː'kuːlə(r)/	n. 预冷器
dual	/'djuːəl/	adj. 双的
approximate	/ə'prɒksɪmət/	v. 接近
appropriate	/ə'prəʊprɪət/	adj. 适当的
pressurization	/ˌpreʃəraɪ'zeɪʃn/	n. 增压
air condition	/eə(r) kən'dɪʃn/	n. 空调
Pressure Regulator and Shutoff Valve (PRSOV)		压力调节与关断活门

Exercise

Fill in the blanks with the following words.

duct　available　obtain　adequate　function　orient

(1) The valve can be opened manually or electrically if power is _____.

(2) Since the instrument was installed, the system is _____ better than before.

(3) The fuel is _____ from the tanks through various valves and ducts.

(4) When the shut off valve is opened, the hydraulic fluid is _____ to the actuators via the flow control valve.

(5) The FMU must be able to provide _____ fuel to the engines according to the pilots' requirement.

(6) The oil system is _____ to provide best circulation of the oil the system.

Supplementary Reading

Pneumatic System

1. General

The pneumatic system supplies hot, high pressure air to the systems on the airplane that use air.

These are the sources of pneumatic power: engine bleed air, APU, and ground source.

2. Engine Bleed Air

There is one bleed air system for each engine. The engine bleed system controls bleed air temperature and pressure.

Engine bleed air comes from the 5th and 9th stages of the high pressure compressor. A high stage regulator and high stage valve control the flow of 9th stage bleed air. The 5th stage check valve prevents reverse flow into the 5th stage.

A bleed air regulator and the pressure regulator and shutoff valve (PRSOV) control the flow of bleed air to the pneumatic manifold.

The air conditioning accessory unit (ACAU) contains relays, and is an interface between the air conditioning/bleed air controls panel switches and lights.

3. Engine Bleed Air Precooler System

The precooler system controls the engine bleed air temperature. The precooler control valve, precooler control valve sensor, and WTAI solenoid valve control the flow of fan air to the precooler.

4. APU Bleed Air

The APU supplies bleed air to the pneumatic manifold. An APU check valve protects the APU from engine bleed air flow.

5. Pneumatic Manifold

The pneumatic manifold gets high pressure air from the source systems and supplies it to the user systems. The pneumatic manifold has these features:

- An isolation valve that can separate the manifold into left and right sides;
- Two duct pressure transmitters (for left and right manifold pressure indication);
- A ground pneumatic connector with a check valve (for an external source of pneumatic power);
- Pressure taps and interfaces for user system valves.

6. Controls and Indications

The air conditioning/bleed air controls panel has control switches and lights to control and monitor the pneumatic system.

Lesson 8   Fuel

A. The airplane fuel system stores and distributes fuel for use by the engines and auxiliary power unit. Many portions of the fuel system are operated automatically by a fuel management system that monitors fuel quantities, fuel distribution in the tanks, B and component status. Fuel system configuration can be displayed on the Engine Indicating and Crew Alerting System (EICAS) by selecting the fuel synoptic. The fuel system can signal fuel system flight conditions and/or faults to the flight crew through EICAS and can record flight conditions and faults in the Central Maintenance Computer System (CMCS) for aid in maintenance.

B. The fuel system is composed of four major subsystems:

1) Storage

The storage sub-system consists of the fuel tanks, a fuel tank ventilation system, and a means to transfer fuel from tank to tank within the airplane. Storage sub-system also includes sealing requirements for the fuel tanks.

2) Distribution

The distribution sub-system consists of the components and tubing necessary to deliver fuel to the engine and auxiliary power unit. Distribution sub-system also includes the refueling and defueling of the fuel tanks.

3) Jettison

The jettison sub-system consists of the components and fuel tubing necessary to jettison fuel overboard through nozzles on the wingtips.

4) Indicating

The indicating sub-system contains components necessary to provide fuel quantity indication by electronic or mechanical means. Quantity measurements determined by the indicating sub-system are used to automatically control fuel feed, transfer, refueling and jettison operations. Indicating sub-system also contains components for indicating low fuel pressure in the fuel feed and jettison sub-system pumps.

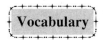
Vocabulary

feed   /fiːd/   v. 供油；喂
Engine Indicating and Crew Alerting System (EICAS)   发动机指示和机组警报系统

Central Maintenance Computer System（CMCS）		中央维护计算机系统
ventilation	/ˌventɪˈleɪʃ(ə)n/	n. 通风
nozzle	/ˈnɒz(ə)l/	n. 喷嘴
vent	/vet/	n. 通气装置
defuel	/diːˈfjuːl/	v. 排油，抽油

Exercise

1. Fill in the blanks with the following words.

fail　due to　lack of　especially　essential　essentially

（1）If the electric power ＿＿＿＿＿＿＿＿, you can use the manual control.

（2）The system has broken down ＿＿＿＿＿＿＿＿ the valve failure.

（3）They will have difficulties in living through the winter due to ＿＿＿＿＿＿＿ enough food and coal.

（4）He is very strong because he likes swimming, ＿＿＿＿＿＿＿ in the autumn when the weather is already quite cold.

（5）We have a lot of work to do, but the ＿＿＿＿＿＿＿ work is to solve the problem in the fueling system.

（6）The system is built up ＿＿＿＿＿＿＿ from the three parts.

（7）The work is a bit difficult for her because she is ＿＿＿＿＿＿＿ experience.

2. Translate the fllowing words into Einglish.

（1）供油系统　　（2）增压泵

（3）内装　　　　（4）防火开关

（5）旁通活门　　（6）交输活门

（7）供油管路　　（8）交输管线

3. Answer the following questions according to the passage.

（1）What is the main purpose of the fuel system?

（2）How many fuel storage parts do the fuel system have?

（3）What are the main functions of the distribution system?

（4）Which systems are the fuel system subdivided into?

（5）What does the defuel system permit?

4. Translate the following sentences into Chinese.

（1）Fueling can be accomplished by the use of a single pressure fueling station through which all tanks can be filled partially or completely.

翻译：_____

_____

(2) However, with use of the fuel boost pump switches, any fuel tank can supply fuel to the APU.

翻译：_____

_____

(3) The center tank boost pumps supply fuel at a higher pressure than the pumps in the main tanks.

翻译：_____

_____

(4) The fueling station is used for defueling and transfer operations which are only possible on the ground.

翻译：_____

_____

(5) Fuel is first supplied to both engines from the center tank and then from the respective tank to engines.

翻译：_____

_____

Supplementary Reading

1. General

The fuel vent system keeps the pressure of the fuel tanks near the ambient pressure. Too large a pressure difference can cause damage to the wing structure.

Drains let fuel in the vent system return to the tanks.

Flame arrestors make sure excessive heat does not enter the fuel vent system. A clogged flame arrestor causes the pressure relief valve in the surge tank to open. When open, the pressure relief valve becomes another vent for the fuel vent system.

2. Component Locations

Stringers and the upper wing skin make the vent channels. The vent channels have drain float valves in the center tank.

Vent tubes attach to vent channels. Each vent tube has a drain float valve.

A fuel vent float valve is on the outboard fuel tank end rib in main tank 1 and main tank 2.

A surge tank drain check valve is on the outboard fuel tank end rib in main tank 1 and main tank 2.

The vent scoop and pressure relief valve are on an access door in each surge tank.

3. Functional Description

Vent channels and vent tubes equalize the pressure between each tank and the surge tanks when the airplane is in a climb attitude. The surge tanks are open to the atmosphere through the vent scoop.

The fuel vent float valves equalize the pressure between main tank 1, main tank 2, and the surge tanks when the airplane is in a cruise or descent attitude.

The surge tank drain check valve permits fuel in the surge tank to flow to either main tank 1 or main tank 2. The surge tank drain check valve also prevents fuel flow from main tank 1 and main tank 2 to the surge tank.

The pressure relief valve prevents damage to the wing structure when there is too much positive or negative pressure in the fuel tanks. The pressure relief valve is usually closed. When closed, it is even with the bottom surface of the wing. When there is too much positive or negative pressure, the pressure relief valve opens. When it is open, part of the pressure relief valve is in the fuel tank. After it opens, the pressure relief valve stays in the open position. In the open position, the pressure relief valve supplies an additional vent in the surge tank. Pull the reset handle to move the pressure relief valve to the closed position.

For normal operations, make sure the pressure relief valve is closed. An open pressure relief valve is a symptom of a problem in the fuel vent system.

Lesson 9   Navigation

Navigation

Description and Operation

A. The navigation systems are those systems used to determine and display the attitude, and position of the airplane with respect to the earth's surface. The navigation systems fall into the following general categories:

(1) The systems which sense and display flight environmental data;

(2) The systems which determine airplane attitude and direction;

(3) The systems which provide landing and taxiing aids;

(4) The systems which are self-contained and independent of ground-based equipment;

(5) The systems which are dependent upon and operate in conjunction with ground-based equipment;

(6) The systems which compute airplane position.

B. The various navigation units are located in four main equipment centers throughout the airplane and are remotely controlled and displayed from the flight compartment. These equipment centers are:

(1) Forward equipment center;

(2) Main equipment center;

(3) Forward center fuselage equipment center;

(4) After center fuselage equipment center.

C. Receiving and/or transmitting antennas required by navigation systems are located externally on the airplane.

D. BEJ 001 – 100

Navigation system information is displayed on the primary flight displays (PFD), navigation displays (ND), and the captain's radio magnetic indicator (RMI). The PFD's and ND's are described in the integrated display system (Ref 31 – 61 – 00). The RMI is described separately under each of the following systems:

(1) Inertial reference system (Ref 34 – 21 – 00);

(2) VOR navigation system (Ref 34 – 51 – 00);

(3) ADF system (Ref 34 – 57 – 00).

Vocabulary

navigation	/ˌnævɪˈgeɪʃ(ə)n/	n. 导航
attitude	/ˈætɪtjuːd/	n. （飞行）姿态
altitude	/ˈæltɪtjuːd/	n. 高度
category	/ˈkætɪg(ə)rɪ/	n. 种类,类别
environmental	/ɪnvaɪrənˈment(ə)l; en-/	adj. 环境的；有关环境的
in conjunction with…		与……一同
various	/ˈveərɪəs/	adj. 各种各样的
antenna	/ænˈtenə/	n. 天线
primary	/ˈpraɪm(ə)rɪ/	adj. 主要的；首要的
separately	/ˈsep(ə)rətlɪ/	adv. 分别地,单独地
ADF(Automatic Direction Finder)		自动定向机
RMI(Radio Magnetic Indicator)		无线电磁指示器
PFD(Primary Flight Displays)		主飞行显示器

Exercise

Translate the following sentences into Chinese.

(1) Antenna severs either or both of the following functions: transmitting or receiving the signals.

翻译：_____

(2) The systems that determine airplane attitude and direction are part of navigation systems.

翻译：_____

(3) The information which is displayed on the PFD is from navigation system.

翻译：_____

Supplementary Reading

ILS Digital Interface

1. General

These are the components that have a digital interface with the multi-mode receivers:

· Captain and first officer NAV control panels;

- Flight data acquisition unit (FDAU);
- Standby attitude indicator;
- Ground proximity warning computer (GPWC);
- Flight management computer (FMC);
- Display electronics units (DEU);
- FCC A and FCC B.

2. Digital Inputs

The NAV control panels supply frequency tune inputs to the multi-mode receivers. The NAV control panels also send tune inputs to the VOR and DME systems on a separate data bus.

3. Digttal Outputs

Each MMR has two output data buses. Output data bus 1 goes to the FCCs. Output data bus 2 goes to many components.

The FDAU receives ILS data and status of the MMR receivers.

The FDAU processes the data for the flight data recorder.

The standby attitude indicator uses localizer and glideslope deviations for the ILS deviation bar operation. Only MMR 1 sends ILS data to the standby attitude indicator.

The GPWC receives glideslope data from both ILS receivers for mode 5 (below glideslope) warnings.

The FMC receives ILS data and receiver condition from the two MMRs. The FMC uses the ILS data for position update calculations.

The FCCs use ILS data to calculate airplane steering commands for the digital flight control system (DECS) autopilot and flight director modes.

The DEU 1 receives two inputs from MMR 1 and two inputs from MMR 2. DEU 2 also receives two inputs from MMR 1 and two inputs from MMR 2. The CDS uses the ILS data to calculate the localizer deviation and glideslope deviation displays.

Lesson 10　Landing Gear

1. General

The landing gear system consists of the gear which support the airplane while on the ground; a gear extension and retraction system; an alternate gear extension system; wheels and brakes for each main gear; means for steering the airplane; gear and door indicating and warning system; control and operation of the gear for landing, take off and ground movement.

2. Main Gear and Doors (Figure 10. 1) (Ref 32 – 10 – 00/001)

1) Main Gear

(1) The main landing gear consists of four main gear, two body gear and two wing gear. Each gear is comprised of a four-wheel truck. The body gear is located at station 1,463.5 in the fuselage. The wing gear is located after of the rear wing spar in board of the engine nacelles at station 1,342.5.

(2) Landing impact is absorbed by five air-oil shock struts, functioning primarily as air springs. Rolling vibrations and variances in runway are absorbed by the hydraulic forces within the shock strut.

(3) Body gear trucks are steerable, providing directional control in conjunction with the nose gear in sharp turns during low speed taxiing and towing. This capability also reduces tire scrubbing in sharp turns.

2) Main Gear Doors

The landing gear doors consist of body gear doors and wing gear doors. The body and wing gear doors each have wheel well doors and shock strut doors. The wheel well doors are hydraulically actuated and can be closed with the gear extended or retracted. The shock strut doors are mechanically attached by linkage rods to the gear shock strut and move only when the gear is moved. All doors are of frame construction with skin paneling on the inner and outer sides. The doors close over all gear openings and fair with the fuselage contour to provide aerodynamic smoothness.

3. Nose Gear and Doors (Ref 32 – 20 – 00/001)

1) Nose Gear

The nose landing gear is a steerable wheel assembly which supports the forward end of the airplane and provides directional control while on the ground. The gear is hydraulically actuated and retracts forward in the wheel well. Landing loads are absorbed by an air-oil shock strut. Steering is provided by hydraulically actuated cylinders. The airplane can be towed forward or after from the nose gear.

2) Nose Gear Door

The nose gear doors are clamshell type and consist of four doors that fair with the fuselage contour when closed. All doors are of alclad rib and skin construction.

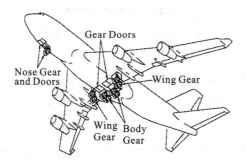

Figure 10.1 Main Gear and Doors

Vocabulary

extension	/ɪkˈstenʃn/	n. 放出
retraction	/rɪˈtrækʃn/	n. 收回
alternate	/ɔːlˈtɜːnət/	adj. 备用
steerable	/ˈstɪərəbl/	adj. 可转弯的
comprised	/kəmˈpraɪzd/	v. 由……组成的
nacelle	/nasɛl/	n. 吊舱
vibration	/vaɪˈbreɪʃn/	n. 振动
variance	/ˈveəriəns/	n. 变化
absorb	/əbˈzɔːb/	v. 吸收
shock strut		减震支柱
sharp turn		急转弯
capability	/ˌkeɪpəˈbɪləti/	n. 能力，容量
scrub	/skrʌb/	n.&v. 摩擦，擦洗
impact	/ˈɪmpækt/	n. 冲击
contour	/ˈkɒntʊə(r)/	n. 轮廓，周线
cylinder	/ˈsɪlɪndə(r)/	n. 筒，圆柱体
fair	/feə(r)/	n. 整流
actuate	/ˈæktʃueɪt/	v. 开动

1. Fill in the blanks with the words given below, one word can only be used once.

support　　　alternate　　　comprised　　　attach

retraction　　　retract　　　actuate

(1) Navigation system is _____ of five sub-systems.

(2) _____ the wire to a hook temporally until the repair is finished.

(3) All the control surfaces are hydraulically _____.

(4) After take-off, the main landing gears are _____ forward into the wheel well.

(5) Landing gears are the main _____ of the aircraft when the aircraft is on the ground.

(6) When the main generators are inoperative, the _____ power supply comes from the main DC buses.

(7) The _____ of the landing gears are actuated by hydraulic power.

2. Answer the following questions according to the passage.

(1) What is the main function of the body landing gears?

_____

(2) What is the main function of the nose landing gear?

_____

(3) What does the main landing gear consist of?

_____

(4) What does the shock struts function as?

_____

(5) How are the shock strut doors attached?

_____

Supplementary Reading

Landing Gear

1. General

The 737 airplane has a tricycle type landing gear with air/oil shock struts.

These are the landing gear structural systems:

- The main landing gear (MLG) and doors (32 - 10);
- The nose landing gear (NLG) and doors (32 - 20).

The landing gear extension and retraction systems extend and retract the main and nose landing gear (32 – 30).

The nose wheel steering system supplies the ground directional control of the airplane (32 – 50).

2. Components

The landing gear system has these main components:
- Control lever assembly;
- Manual extension mechanism;
- Transfer valve;
- Selector valve;
- Main landing gear (2);
- Nose landing gear;
- Shimmy damper;
- Proximity switch electronics unit (PSEU);
- Landing gear panel;
- Auxiliary landing gear position lights.

3. General Description

Hydraulic system A normally supplies pressure to the landing gear extension and retraction. Hydraulic system B supplies pressure for retraction only.

The landing gear transfer valve receives electrical signals from the proximity switch electronics unit (PSEU). The landing gear transfer valve changes the pressure source of the landing gear from hydraulic system A to hydraulic system B.

See the Air/Ground System section for more information about the proximity switch electronics unit (PSEU) (SECTION 32 – 09).

You move the landing gear control lever assembly to control landing gear extension and retraction. The control lever moves the selector valve through cables.

The selector valve also gets an electrical input from the manual extension system. This operates a bypass valve in the selector valve to connect the landing gear retraction to the hydraulic system return. This lets the manual extension system extend the landing gear.

Landing gear lights show the position of the landing gear. The PSEU receives landing position signals from sensors on the landing gear. The normal and auxiliary lights are controlled by the PSEU.

Lesson 11　Engine

The CFM56-7 is a high bypass, dual rotor, axial flow turbofan engine. The engine fan diameter is 61 inches (1.55 meters). The bare engine weight is 5,257 pounds (2,385 kilograms).

The engine has these sections:

Fan and booster, High pressure compressor (HPC), Combustor, High pressure turbine (HPT), Low pressure turbine (LPT), Accessory drive.

The fan and booster rotor and the LPT rotor are on the same low pressure shaft (N1).

The HPC rotor and the HPT rotor are on the same high pressure shaft (N2).

1. Fan and Booster

The fan and booster is a four-stage compressor. The fan increases the speed of the air. A splitter fairing divides the air into these two air flows: primary and secondary.

The primary air flow goes into the core of the engine. The booster increases the pressure of this air and sends it to the HPC. The secondary air flow goes in the fan duct. It supplies approximately 80 percent of the thrust during take-off.

2. High Pressure Compressor (HPC)

The HPC is a nine-stage compressor. It increases the pressure of the air from the LPC and sends it to the combustor. The HPC also supplies bleed air for the aircraft pneumatic system and the engine air system.

3. Combustor

The combustor mixes air from the compressors and fuel from the fuel nozzles. This mixture of air and fuel burns in the combustion chamber to make hot gases. The hot gases go to the HPT.

See the engine fuel and control chapter for more information on the fuel nozzles. (CHAPTER 73)

4. High Pressure Turbine (HPT)

The HPT is a single-stage turbine. It changes the energy of the hot gases into a mechanical energy. The HPT uses this mechanical energy to turn the HPC rotor and the accessory drive.

5. Low Pressure Turbine (LPT)

The LPT is a four-stage turbine. It changes the energy of the hot gases into a mechanical energy. The LPT uses this mechanical energy to turn the fan and booster rotor.

6. Accessory Drive

The accessory drive has these components：

- Inlet gear box (IGB)；
- Radial drive shaft (RDS)；
- Transfer gear box (TGB)；
- Horizontal drive shaft (HDS)；
- Accessory gear box (AGB).

The N2 shaft turns the AGB through these shafts and gearboxes：

- IGB；
- RDS；
- TGB；
- HDS.

The AGB holds and operates the airplane accessories and the engine accessories. They are described later in this section.

Vocabulary

engine	/ˈendʒɪn/	*n*. 发动机
high pressure compressor		高压压气机
combustor	/kəmˈbʌstə/	*n*. 燃烧室
turbine	/ˈtɜːbaɪn/	*n*. 涡轮
inlet gear box		内部齿轮箱
radial drive shaft		径向传动轴
transfer gear box		转换齿轮箱
accessories	/əkˈsɛsəriz/	*n*. 附件

Exercise

1. Please fill in the blanks with the following words, and, if necessary, change the grammatical form.

complete	until	reference	supply	injury
adjacent to	access	align	move	cause

（1）If you are not sure about what to do, a _____ to the MM may help you.

（2）You must remove the panel to get _____ to the part.

（3）Keep your seatbelt fastened _____ the airplane has reached its final parking position.

（4）During the test persons must be clear of all control surfaces because they can _____ quickly.

（5）If you do not read the WARNING carefully, personal _____ may be the result.

（6）Improper handling will _____ damage to the equipment.

（7）Don't forget to _____ electrical power before you start the test.

（8）Parking position V146 is _____ V147.

（9）When the test is _____ you must put the airplane back to its usual condition.

（10）He _____ his papers in geometrical order on his desk.

2. Answer "yes" or "no" and, if necessary, correct the statements below.

（1）During the yaw damper actuator test the elevator moves in the left and right directions. （T/F）

（2）When the test is completed the INOP light goes off. （T/F）

（3）Before you align the IRUS in NAV mode you must supply electrical power. （T/F）

（4）The presence of personnel and equipment is permitted near the control surfaces before supplying hydraulic power. （T/F）

（5）The FLT CONTROL HYD POWER TAIL switches are on the copilot's Overhaul Maintenance Panel. （T/F）

（6）To prepare the upper yaw damper for the test you must push and release the engage switch of the upper yaw damper on panel P5. （T/F）

（7）After that the INOP annunciator must be on. （T/F）

（8）If ＜Return shows after you push the LSK, push the NEXT PAGE key. （T/F）

（9）If INHIBITED shows, the test will not operate. （T/F）

（10）To prepare the CDU for the test you must push the MENU key first. （T/F）

Supplementary Reading

Engine—Main Engine Bearing

1. General

Five main engine bearings hold the N1 shaft and the N2 shaft. Numbers from 1 to 5 identify the engine bearings. Ball bearings absorb the axial and the radial loads from the shafts. Roller bearings absorb only radial loads.

The main engine bearings are in two sump cavities. The sump cavities are the forward sump and the rear sump.

2. Main Engine Bearings

The number 1 and the number 2 bearings hold the front of the N1 shaft.

One ball bearing and one roller bearing are the number 3 bearing assembly. Both number 3 bearings hold the front of the N2 shaft.

The number 4 bearing holds the rear or the N2 shaft.

The number 5 bearing holds the rear of the N1 shaft.

The number 1, 2 and 3 bearings are in the forward sump. The number 4 and 5 bearings are in the rear sump.

Engine—Engine Aerodynamic Stations

General

There are probes or sensors at these 5 aerodynamic stations on the CFM56—7.

Station 0 (ambient air);

Station 12 (fan inlet);

Station 25 (high pressure compressor inlet);

Station 30 (high pressure compressor discharge);

Station 49. 5 (stage 2 low pressure turbine stator).

See the engine fuel and control chapter for more information on engine fuel and control (SECTION 73 – 00).

See the engine indicating chapter for more information on the probes and sensors (SECTION 73 – 21).

Lesson 12　Flight Control

1. General

The pilots manually operate the flight controls through cables. The autopilot automatically operates them.

2. Aileron

The aileron control wheels move cables that give input to the aileron feel and centering unit. This controls the aileron PCUs. The PCUs move the aileron wing cables and the ailerons.

The aileron trim switches give an input to the aileron feel and centering unit and change the neutral position of the ailerons.

Aileron PCD movement also goes to the spoiler mixer. The mixer moves the flight spoiler wing cables, which control the flight spoiler actuators. The actuators move the flight spoilers to assist the ailerons for roll control.

The autopilot actuators give a mechanical input to the PCUs through the feel and centering unit. The PCUs move the aileron wing cables and the ailerons and they give input to the spoiler mixer.

3. Elevator

The control columns move cables that give input to the elevator feel and centering unit. This controls the elevator PCUs. The PCUs move torque tubes that move the elevators.

The autopilot actuators give a mechanical input to the PCUs through the feel and centering unit. The PCU moves the elevators.

4. Rudder

The rudder pedals move cables that give an input to the rudder feel and centering unit. This controls the rudder PCUs. The rudder PCUs move the rudder.

The rudder trim switch gives an input to the rudder feel and centering unit and change the neutral position of the rudder.

5. Flaps and Slats

The flap control lever moves the trailing edge flap control valve. Hydraulic pressure goes through the valve and drives the hydraulic motor. The hydraulic motor supplies power to the flap drive system and the flaps move. Follow-up cables give feedback to the trailing edge flap control valve to stop the laps at the commanded position.

The follow-up cables also give an input to the leading edge flaps control valve. This controls the position of the leading edge devices. Then hydraulic pressure goes to the actuators and moves the leading edge flaps and slats.

The alternate flap switches electrically control the trailing edge flaps. They also control the standby hydraulic system to extend the leading edge flaps and slats.

6. Spoilers and Speedbrakes

The speedbrake lever moves cables that control the spoiler mixer. The mixer moves the spoiler wing cables that control the flight spoiler actuators. The mixer also moves the ground spoiler control valve. The valve supplies hydraulic pressure to the ground spoiler actuators to raise the spoilers.

During automatic deployment, the auto speedbrake actuator gives input to the same cables as above and back drives the speedbrake lever.

7. Horizontal Stabilizer

The stabilizer trim wheels move cables that give an input to the gearbox. The gearbox moves a jackscrew and moves the stabilizer.

The electric stabilizer trim switches control an electric motor near the gearbox. The motor moves the gears to move the stabilizer. The autopilot also controls the stabilizer trim motor.

When the stabilizer moves, it also moves the elevators through the elevator tee 1 and centering unit.

Vocabulary

aileron	/ˈeɪlərɒn/	n. 副翼
elevator	/ˈelɪveɪtə(r)/	n. 升降舵
rudder	/ˈrʌdə(r)/	n. 方向舵
flap	/flæp/	n. 襟翼
slat	/slæt/	n. 缝翼
spoiler	/ˈspɔɪlə(r)/	n. 扰流板
horizontal stabilizer		水平安定面

trim	/trɪm/	v. 配平

1. Please fill in the blanks with the following words.

performance	previously	operate	fall	actual
accuracy	encountered	failure	available	respond to

(1) The _____ of engine 2 caused the accident.

(2) If the indicators _____ to illuminate, the CADC must be replaced.

(3) Electrical power must be _____, otherwise the operational test can't be accomplished.

(4) The _____ condition of the system couldn't be estimated at first sight.

(5) Absolute _____ is required for performing the test successfully.

(6) If faults are _____, rectify them as soon as possible.

(7) The air data instruments must _____ input signals.

(8) The accurate _____ of maintenance procedures is essential.

(9) When all steps in the OK condition path check out, the CADC system is _____.

(10) You have to verify that the altimeters show the _____ set altitude.

2. Find out the following statements are true or false, and, if necessary, correct them.

(1) When a test step does not check out, follow the NOT OK line. (T/F)

(2) All trouble-shooting procedures are based on the assumption that electrical power is available. (T/F)

(3) Prepare for trouble shooting. Verify that auto throttle systems are on. (T/F)

(4) Assure probe heaters are off. (T/F)

(5) If one or more yellow annunciator lights are lit: Press and hold SELF TEST or button A switch. (T/F)

(6) If all indicators do not come on: Replace lamps in faulty indicators. (T/F)

(7) If one indicator did not slew or flag did not appear: Replace CADC immediately. (T/F)

(8) If warning lamp conditions are not satisfied: Replace CADC and perform specified test. (T/F)

(9) If altimeter indication is incorrect: Refer to procedure 2. (T/F)

(10) If TAT indication is incorrect: Replace TAT indicator and do specified tests. (T/F)

Flight Controls Multiple-Use System/Units

1. General

The flight controls system has some multiple use components. This section supplies information for these multiple use components:

- Flight control cables;
- Flight control panel;
- Flight control hydraulic modular packages.

2. Flight Control Cables

You use the flight control cables to give manual input to each flight control system.

3. Flight Control Panel

The flight control panel has hydraulic control switches and caution lights for several of the flight control systems.

Flight Control Hydraulic Modular Packages

The two flight controls hydraulic modular packages control and monitor hydraulic pressure to the flight controls. Each package contains these components:

- Flight controls shutoff valve;
- Low pressure warning switch;
- Spoiler shutoff valve;
- Compensator cartridge.

1. General

This is the system A and B hydraulic distribution to the primary flight controls. The pressure goes to the flight control hydraulic modular package. Then it goes through the flight controls shutoff valve. A low pressure warning switch monitors for low pressure to the flight controls. Then the pressure goes to the using components.

2. Purpose

The flight control cables give manual input to each flight control system.

3. Location

Most cables are under the floor and go from the flight compartment to the respective flight controls.

4. Physical Description

The cables are steel components specially treated for corrosion protection.

5. Functional Description

Each cable set is a pair. During a normal input, the tension in one cable increases and moves the components downstream. These are the pairs of cables:

- Aileron control bus cables (ACBA, ACBB);
- Aileron left and right body cables (AA, AB);
- Aileron wing cables (ABSA, ABSB);
- Elevator control cables (EA, EB);
- Rudder control cables (RA, RB);
- Flap control cables (WFA, WFB);
- Flap follow-up cables (WFFA, WFFB);
- Stabilizer control cables (STA, STB);
- Speedbrake control cables (SBA, SBB);
- Spoiler control cables (WSA, WSB).

Lesson 13   Oxygen

1. Purpose

The oxygen systems supply oxygen to the flight crew, cabin attendants and passengers. Oxygen can be used for most reasons:

- Life sustaining oxygen if the plane depressurizes;
- Emergencies;
- First aid.

2. General Description

The flight crew oxygen system operates independently of the other systems. It is a high pressure gaseous system. High pressure gaseous oxygen is in a cylinder in the EE compartment. The manifold supplies oxygen to the flight crew oxygen masks.

The passenger oxygen system uses chemical oxygen generators. The generators are in the passenger service units (PSUs). Each chemical generator is separate, and supplies only its masks. The masks connect to the chemical generators by flexible tubes.

3. Crew Oxygen Purpose

The flight crew oxygen system supplies the flight crew with low pressure gaseous oxygen.

4. General Description

High pressure gaseous oxygen is stored in a cylinder.

Over pressure devices protect the cylinder. A green plastic discharge indication disk on the fuselage skin shows cylinder discharge from overpressure (when the disk is missing).

A cylinder head assembly connects the cylinder to the airplane distribution system.

The oxygen supply lines are made of seamless stainless-steel tubing and use flareless fittings.

The flight crew masks supply the oxygen to the crew. The masks have diluter-demand regulators and controls. The masks are modular, independently adjustable, and easy to put on.

When cylinder pressure is too low for operational requirements, you replace it.

5. Location

These system components are in the EE compartment：

- High pressure oxygen cylinder；
- Cylinder head assembly；
- Overpressure discharge tubing.

6. Training Information Point

Oxygen system maintenance requires special care and cleanliness. Oxygen system maintenance personnel should know the special materials and procedures used in system servicing. Refer to chapter 12 of the maintenance G manual for these materials and procedures.

Keep oxygen systems clean and dry. Use approved cleaning materials.

<u>WARNING</u>：DO NOT ALLOW OIL. GREASE, DIRT OR OTHER FLAMMABLE MATERIALS TO TOUCH OXYGEN SYSTEM COMPONENTS. THESE MATERIALS WHEN EXPOSED TO PRESSURIZED OXYGEN CAN IGNITE AND CAUSE AN EXPLOSION. A FIRE OR EXPLOSION CAN CAUSE INJURY TO PERSONS AND DAMAGE TO EQUIPMENT.

Vocabulary

oxygen	/ˈɒksɪdʒən/	n. 氧气
first aid		医疗救生
mask	/mɑːsk/	n. 面罩
flexible tube		软管
gaseous oxygen		气态氧
overpressure discharge tubing		过压释放管

Exercise

1. Find the words or phrases which can match the words or phrases given in the left column.

in conjunction with	make sure
perform	block
ensure	describe in detail
mount	allow

damp   together with

permit   next to

adjacent to   do

elaborate   install

2. Fill in the blanks with the words given below with proper form.

Adjacent   fix perform   actuate   in conjunction   with   avoid   demand

(1) If system A fails, system B, _____ standby system, will actuate the control surfaces.

(2) All the flight control surfaces _____ by hydraulic power.

(3) There is a large _____ of electricity power for a four engine wide body aircraft.

(4) Be careful when removing this part, not to damage the _____ system.

(5) _____ the wiring before you remove the nut.

(6) This repair should _____ with great care.

(7) Always _____ using solvent in a badly ventilated area.

┌─────────────────────────┐
│ **Supplementary Reading** │
└─────────────────────────┘

Oxygen-Passenger

1. Purpose

　　The passenger oxygen system supplies emergency oxygen to the passengers and cabin attendants.

2. Location

　　Passenger oxygen generators, masks, firing pin mechanisms and deployment door latch actuators are in these units:

　　• Passenger service units (PSUs);

　　• Lavatory service units (LSUs);

　　• Attendant service units (ASUs).

　　A guarded toggle switch for manual release of the passenger oxygen masks is on the P5 aft overhead panel.

　　A pressure switch for automatic release of the passenger oxygen mask is in the J23 junction box in the EE compartment.

3. General Description

The passenger oxygen system uses chemical generators to make oxygen. Oxygen from the generators flows through flexible supply hoses to the passenger oxygen masks.

Passenger oxygen masks deploy electrically one of these two ways:

• Manually by the crew with a guarded toggle switch on the oxygen system control panel (P5);

• Automatically by operation of a pressure switch (14,000 feet cabin altitude).

4. Operational Displays

The PASS OXY ON light on the P5 after overhead panel comes on when the passenger oxygen masks deploy.

Section 3

Gas Turbine Engines

Lesson 14   Conspectus of a Gas Turbine Engines

Engine Operation

The front orinlet duct is almost entirely open to permit outside air to enter the front of the engine. The compressor works on the incoming air and delivers it to the combustion section with as much as twelve times or more the pressure the air had when it entered the engine. In the combustion section fuel is sprayed and mixed with the compressed air. The air-fuel mixture is then ignited by igniter or spark plugs. Once the mixture is lighted the igniter is turned off as the burning process will continue without further assistance, providing the engine is supplied with the right air-fuel mixture. This air-fuel mixture bums at a relatively constant pressure with only about 25 percent of the air taking part in the actual combustion process the rest of the air is used for cooling some of the components of the combustion section. This heated expanding air is then directed to the turbine. The turbines are connected to the compressor and accessories. They will extract a major portion (nearly two-thirds) of the energy to drive them. The remaining energy will be used to either propel the aircraft through a jet nozzle or to turn other times (power turbines).

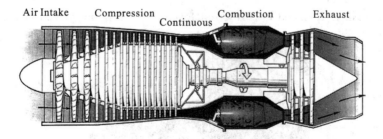

Figure 14. 1   Gas Turbine Engines

Power turbines will take this energy and transform it into shaft horsepower, which in turn drives an output shaft or propeller from a power shaft (Figure 14. 1).

Types of Gas Turbine Engines

The gas turbine engines most commonly used today are divided into four types.

Turbojet engine—a gas turbine engine in which the net energy available is used by the air or hot gas solely in the form of the jet issuing through a propelling nozzles or a jet engine whose air is supplied by a turbine—driven compressor, the turbine being driven by exhaust gases.

Turboshaft engine—an engine in which the net energy available is transmitted from the exhaust gas-driven turbine wheel through a planetary gear, which may be integral to the engine or externally provided by the aircraft manufacturer, to turn ahelicoptertransmission-driven powertrain system.

Turboprop engine (Figure 14.2)—a turboshaft engine in which power shaft through reduction gearing to a propeller. The major difference between the two engines is the gear reduction.

Figure 14.2　Turboprop Engine

Turbofan engine (Figure 14.3)—in principle, the same as a turboprop, except that the geared propeller is replaced by a duct-enclosed axial-flow fan driven at engine speed.

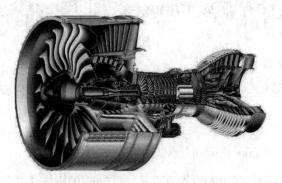

Figure 14.3　Turbofan Engine

The term "turbo" means "turbine". Therefore, a turboshaft engine is an engine that delivers power through power through a shaft, which, in turn is powered by a turbine wheel. Army aircraft gas turbine engines are generally the free-type power turbine turboprop or turboshaft engines. Having a free power turbine enables the power output shaft to turn at a constant speed while the power-producing capability of the engine can be varied to accommodate the increased loads applied to the power output shaft. Turbine engines may also be classified into two general groups，centrifugal-flow and axial-flow，depending on the type of compressor used. However，most gas turbine engines in Army aircraft employ a combination of both types.

Vocabulary

inlet	/ˈɪnlet/	n. 进气道
compressor	/kəmˈpresə(r)/	n. 压气机
combustion	/kəmˈbʌstʃən/	n. 燃烧
spray	/spreɪ/	v. 飞溅
mixture	/ˈmɪkstʃə(r)/	n. 混合物
burning	/ˈbɜːnɪŋ/	v. 燃烧
gas turbine engines		燃气涡轮发动机
turbojet engine		涡喷发动机
turboshaft engine		涡轴发动机
turboprop engine		涡桨发动机
turbofan engine		涡扇发动机

Exercise

Complete the following sentences and translate them into Chinese.

(1) The compressor works on the _____ and delivers it to the _____ with as much as twelve times or more the pressure the air had when it entered the engine.

翻译：_____

_____

(2) In the combustion section fuel is _____ .

翻译：_____

_____

(3) The turbines are connected to _____ .

翻译：_____

_____

（4）Turbofan engine—in principle, the same as a turboprop, except that the geared propeller is replaced by _____.

翻译：_____

_____

（5）What are the types of gas turbine engines, _____

_____.

翻译：_____

_____

（6）Army aircraft gas turbine engines are generally _____.

翻译：_____

_____

Supplementary Reading

Jet engines move the airplane forward with a great force that is produced by a tremendous thrust and causes the plane to fly very fast.

All jet engines, which are also called gas turbines, work on the same principle. The engine sucks air in at the front with a fan. A compressor raises the pressure of the air. The compressor is made up of fans with many blades and attached to a shaft. The blades compress the air. The compressed air is then sprayed with fuel and an electric spark lights the mixture. The burning gases expand and blast out through the nozzle, at the back of the engine. As the jets of gas shoot backward, the engine and the aircraft are thrust forward (Figure 14. 4).

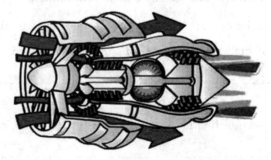

Figure 14. 4   How the Air Flows Through a Jet Engine

The image above shows how the air flows through the engine. The air goes through the core of the engine as well as around the core. This causes some of the air to be very hot and some to be cooler. The cooler air then mixes with the hot air at the engine exit area.

A jet engine operates on the application of Sir Isaac Newton's third law of physics: for every action there is an equal and opposite reaction. This is called thrust. This law is demonstrated in simple terms by releasing an inflated balloon and watching the escaping air propel the balloon in the opposite direction. In the basic turbojet engine, air enters the front intake and is compressed, then forced into combustion chambers where fuel is sprayed into it and the mixture is ignited. Gases which form expand rapidly and are exhausted through the rear of the combustion chambers. These gases exert equal force in all directions, providing forward thrust as they escape to the rear. As the gases leave the engine, they pass through a fan-like set of blades (turbine) which rotates the turbine shaft. This shaft, in turn, rotates the compressor, thereby bringing in a fresh supply of air through the intake. Engine thrust may be increased by the addition of an afterburner section in which extra fuel is sprayed into the exhausting gases which burn to give the added thrust. At approximately 400 m/h, one pound of thrust equals one horsepower, but at higher speeds this ratio increases and a pound of thrust is greater than one horsepower. At speeds of less than 400 m/h, this ratio decreases.

In a turboprop engine, the exhaust gases are also used to rotate a propeller attached to the turbine shaft for increased fuel economy at lower altitudes. A turbofan engine incorporates a fan to produce additional thrust, supplementing that created by the basic turbojet engine, for greater efficiency at high altitudes. The advantages of jet engines over piston engines include lighter weight with greater power, simpler construction and maintenance with fewer moving parts, and efficient operation with cheaper fuel.

Lesson 15　Inlet

The amount of intake required by a gas turbine engine is approximately 10 times that required by a reciprocating engine. The air entrance is designed to conduct incoming air to the compressor with minimum energy loss resulting from drag or ram pressure loss, that is, the flow of air into the compressor should be free of turbulence to achieve maximum operating efficiency. Proper design contributes materially to aircraft performance by increasing the ratio of compressor discharge pressure to duct inlet pressure.

The amount of air passing through the engine depends on the below:

- Speed of the compressor RPM;
- Forward speed of the aircraft;
- Density of the ambient air.

Inlets (Figure 15. 1) are classified as:

- Nose inlets—located in the nose of the fuselage, power plant Pod, or nacelle.

- Wing inlets—located along the leading edge of the wing usually at the root for single engine installations.

- Annular inlets—encircling in whole or in part the fuselage or power plant pod or nacelle.

- Scoop inlets—projecting beyond the immediate surface of the fuselage or nacelle.

- Flush inlets—recessed in the side of the fuselage, power plant or nacelle.

- Bellmouth inlets—a bell-shaped funnel with carefully rounded shoulders, mounted to front of engine.

Figure 15. 1　Types of Inlet

There are two basic types of air entrances in use: single entrance and divided entrance. Generally, it is best to use a single entrance with an axial-flow engine to obtain maximum ram pressure through straight flow. Single entrance is used almost exclusively on wing or external installations where the unobstructed entrance lends itself readily to a single short, straight duct.

A divided entrance offers greater opportunity to diffuse the incoming air and enter the plenum chamber with the low velocity required to utilize efficiently a double-entry compressor. (The plenum chamber is a storage place for ram air, usually associated with fuselage installation.) It is also advantageous when the equipment installation or pilot location makes the use of a single or straight duct impractical. Inmost cases the divided entrance permits the use of very short ducts with a resultant small pressure drop through skin fiction. The air inlet section of turboprop and turboshaft engines also incorporates some type of particle separator or inlet screens to protect compressors from foreign object damage (FOD). Systems will vary among manufactures. Consult the aircraft or engine technical manuals for a description of the inlet duct and. Its particular air inlet protective device.

Air inlet ducts have an anti-icing system. Turbine engine air inlets use hot engine oil, hot bleed air, or a combination of both these systems prevent icing in a turbine engine air inlet. Then are not designed to melt ice that has already formed on or in the inlets.

Guide vanes are included in some turbine engines these wanes direct air coming through the inlet into the compressor at the most efficient angle. The angle depends on the speed of the engine. On most engines the vanes are hollow to allow hot air or oil to flow through to prevent ice build up.

Vocabulary

the amount of…		……的数量
turbine	/ˈtɜbaɪn/	n. 涡轮
approximately	/əˈprɒksɪmətli/	adv. 大约
reciprocating	/rɪˈsɪprəˌkeɪtɪŋ/	v. 往复
entrance	/(for v.)ɪnˈtrɑːns; (for n.) ˈentrəns/	n. 入口
conduct	/kənˈdʌkt; ˈkɒndʌkt/	v. 实施
minimum	/ˈmɪnɪməm/	adj. 最小的
located in…		位于……
fuselage	/ˈfjuzəlɑʒ/	n. 机身

| wing | /wɪŋ/ | *n.* 机翼 |
| annular | /ˈænjələ(r)/ | *adj.* 环形的 |

Exercise

Complete the following sentences and translate them into Chinese.

（1）The air entrance is designed to conduct incoming air to the compressor with _____, that is, the flow of air into the compressor should _____.

翻译：_____

_____

（2）Annular inlets—encircling in _____.

翻译：_____

_____

（3）There are two basic types of air entrances in use: single entrance and divided entrance.

翻译：_____

_____

（4）Single entrance is used almost exclusively on wing or external installations where the unobstructed entrance lends itself readily to a single short, straight duct.

翻译：_____

_____

（5）In most cases the divided entrance permits the use of very short ducts with a resultant small pressure drop through skin fiction.

翻译：_____

_____

（6）Turbine engine air inlets use hot engine oil, hot bleed air, or a combination of both these systems prevent icing in a turbine engine air inlet.

翻译：_____

_____

（7）Guide vanes are included in some turbine engines these wanes direct air coming through the inlet into the compressor at the most efficient angle.

翻译：_____

_____

Propulsion is the science of designing an engine to propel a vehicle forward or up. For aviation, propulsion is generally broken into 2 categories: air-breathing propulsion for airplanes and rocket propulsion for spacecraft. Both work on the principle of pushing high velocity exhaust gases out the back end (reaction thrust principle), but they differ in one significant detail. An air-breathing engine uses the air steam in which the airplane is flying to augment the propulsive abilities of the engine so it can carry less fuel. A rocket engine travels in space where there is no air, and therefore it must carry all its fuel internally. An air-breathing engine will have both an inlet and an exit, while the rocket will be closed in the front and an exit, while the rocket will be closed in the front and only have an exit. In general, an air-breathing engine will get more thrust for less fuel than a rocket.

The forces of flight (lift, drag, weight and thrust) were discussed in the lesson 3. Thrust is the forward facing force generated by the engines of the airplane. The air flows into the engine at roughly the flight speed of the airplane, and it exits the engine flowing much hotter and faster. The thrust is computed using the rate of mass flowing through the engine times the difference between the high velocity of the exhaust gases and the original velocity of the air into the inlet. The exhaust gases flow out the back of the engine, causing a reaction force on the airplane, pushing it forward. This concept is called the reaction thrust principle.

Lesson 16　Compressor

The compressor section of the turbine engine has many functions. Its primary function is to supply enough air to satisfy the requirements of the combustion burners. The compressor must increase the pressure of the mass of air received from the air inlet duct and then discharge it to the burners in the required quantity and pressure.

A secondary function of the compressor is to supply bleed air for various purposes in the engine and aircraft. The bleed air is taken nom any of the various pressure stages of the compressor. The exact location of the bleed port depends on the pressure or temperature required for a particular job. The ports are small openings in the compressor case adjacent to the particular stages from which the air is to be bled. Varying degrees of pressure and heat are available simply by tapping into the appropriate stage. Air often bled from the final or highest-pressure stage because at this point pressure and air temperature are at a maximum. At time it may be necessary to cool this high-pressure air. If it is used for cabin pressurization or other purposes where excess heat would be uncomfortable or detrimental the air is sent through a refrigeration unit.

Bleed air has various uses including driving the remote-driven accessories. Some current applications of bleed air are:

- In cabin pressurization heating and cooling;
- In deicing and anti-icing equipment;
- For pneumatic starting of engines;
- In auxiliary drive units (ADUs);
- In control booster servo systems;
- As power for running instruments.

Compressor section location depends on the type of compressor. In the centrifugal-flow engine the compressor is between the accessory section and the combustion section; in the axial-flow engine the compressor is between the air inlet duct and the combustion section.

1. Centrifugal-Flow Compressor (Figure 16. 1)

The centrifugal-flow compressor basically consists of an impeller (rotor), a diffuser (stator) and a compressor manifold. The impeller and the diffuser are the two main functional elements. Although the diffuser is a separate component positioned inside and secured to the manifold, the entire assembly (diffuser and manifold) is often referred to as the diffuser.

The impeller's function is to pick up and accelerate air outward to the diffuser.

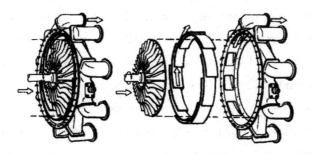

Figure 16. 1　Centrifugal-Flow Compressor

Impellers may be either of two types—single entry or double Entry. Both are similar in construction to the reciprocating engine supercharger impeller. The double-entry type is similar to two back-to-back impellers. However, because of much greater combustion air requirements in turbine, engines, these impellers are larger than supercharger impellers.

The principal differences between the two types of impellers are size and ducting arrangement. The double-entry type has a smaller diameter but is usually operated at a higher rotational speed to ensure enough airflow. The single-entry impeller permits convenient ducting directly to the impeller eye (inducer vanes) as opposed to the more complicated ducting necessary to reach the rear side of the double-entry type. Although slightly more efficient in receiving air, the single-entry impeller must be large in diameter to deliver the same quantity of air as the double-entry type. This of course, increases the overall diameter of the engine. Included in the ducting for double-entry compressor engines is the plenum chamber. This chamber is necessary for a double-entry compressor because air must enter the engine at almost right angles to the engine axis. To give a positive flow, air must surround the engine compressor at a positive pressure before entering the compressor.

Multistage centrifugal compressors consist of two or more single compressors mounted in tandem on the same shaft. The air compressed in the first stage passes to the second stage at its point of entry near the hub. This stage will further compress the air and pass it to the next stage if there is one. The problem with this type of compression is in turning the air as it is passed from one stage to the next.

The diffuser is an annular chamber provide with a number of vanes forming a series of divergent passages into the manifold. The diffuser vanes direct the flow of air from the impeller to the manifold at an angle designed to retain the maximum amount of energy imparted by the impeller. They also deliver the air to the manifold at a velocity and pressure satisfactory for combustion chambers.

The compressor manifold diverts the flow of air from the which, which is an integral part of the manifold, into the combustion chambers. The manifold will have one outlet port for each chamber so that the air is evenly divided. A compressor outlet elbow is bolted to each of the outlet ports. These air outlets are constructed in the form of ducts and are known by a variety of names including "air outlet ducts", "outlet elbows", and "combustion chamber inlet ducts". These outlet ducts perform a very important part of the diffusion process. They change the airflow direction form radial to axial. The diffusion process is completed after the turn to help the elbows perform this function efficiently, turning vanes (cascade vanes) are sometimes fitted inside the elbows. The vanes reduce air pressure losses by presenting a smooth, turning surface.

The centrifugal compressor is used best on smaller engines where simplicity flexibility, and ruggedness are primary requirements. These have a small frontal area and can handle high airflows and pressures with low of efficiency.

Centrifugal-flow compressors have the following advantages:

- High pressure rise per stage;
- Efficiency over wide rotational speed range;
- Simplicity of manufacture with resulting low cost;
- Low weight;
- Low starting power requirements.

They have the following disadvantages:

- Large frontal area for given airflow;
- Impracticality if more than two stages because of losses in turn between stages.

2. Axial-Flow Compressor (Figure 16. 2)

Axial-flow compressor have two main elements: a rotor (drum or disc type) and a stator. These compressors are constructed of several different materials depending on the load and operating temperature. The drum-type rotor consists of rings that are flanged to fit one against the other so that the entire assembly can be held together by through bolts. This type of construction is satisfactory for low-speed compressors where centrifugal stresses are low. The rotor (disc-type) assembly consists of the following:

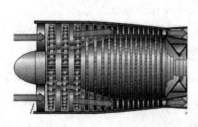

Figure 16. 2   Axial-Flow Compressors

- Stub shafts;
- Discs, Blades;
- Ducts;
- Air vortex spoilers;
- Air vortex spoilevs;
- Spacers;
- Tie bolts;
- Torque cones.

Rotor blades are generally machined from stainless steel forgings, although some may be made of titanium in the forward (colder) section of the compressor. The blades are attached in the disc rim by different methods using either the fir-tree-type, dovetail-type, or bulb-type root designs. The blades are then locked into place with screws, peening, locking wires, pins, keys or plates. The blades do not have to fit too tightly in the disc because centrifugal force during engine operation causes them to seat. Allowing the blades some movement reduces the vibrational stresses produced by high-velocity airstreams between the blades. The newest advance in technology is a one-piece design machined blade disc (combined disc and blade); both disc and rotor blade are forged and then machined into one.

Clearances between rotor blades and the outer case are important to maintain high efficiency. Because of this, some manufacturers use a "wear fit" design between the blade and outer case. Some companies design blades with knife-edge tips that wear away to form their own clearances as they expand form the heat generated by air compression. Other companies coat the inner surface of the compressor case with a soft material (Teflon) that can be worn away without damaging the blade. Rotor discs that are joined together by tie bolts use serration lines or curve coupling teeth to prevent the discs from turning in relation to each other. Another method of joining rotor discs is at their rims.

Axial-flow compressor casings not only support stator vanes and provide the outer wall of the axial paths the air follows but also provide the means for extracting compressor air for various purposes. The stator and compressor cases show great differences in design and construction. Some compressor cases have variable stator vanes as an additional feature.

Other (compressor cases) have fixed stators. Stators vanes may be either solid or hollow and mayor may not be connected at their tips by a shroud. The shroud serves two purposes. First, it provides support for the longer stator vanes located in the forward stages of the compressor. Second, it provides the absolutely necessary air seal between rotating and stationary parts. Some manufacturers use split compressor cases while others favor a weldment, which forms a continuous case. The advantage of the split case is that the compressor and stator blades are readily available for inspection or maintenance. On the other hand, the continuous case offer simplicity and strength since it requires no vertical or horizontal parting surface.

Both the case and the rotor are very highly stressed parts. Since the compressor turns at very high speed the discs must be able to withstand very high centrifugal forces. In addition, the blades must resist bending loads and high temperatures. When the compressor is constructed, each stage is balanced as a unit. The compressor case in most instances is one of the principals of aluminum steel or magnesium.

Axial-flow compressors have the following advantages:

- High peak efficiency;
- Small frontal area for given airflow;
- Straight-through flow, allowing high ram efficiency;
- Increased pressure rises due to increased number of stages with negligible losses.

They have the following disadvantages:

- Good efficiency over narrow rotational speed range;
- Difficulty of manufacture and high cost;
- Relatively high weight;
- High starting power requirements (this has been partially overcome by split compressors).

The air in an axial compressor flows in an axial direction through a series of rotating (rotor) blades and stationary (stator) vanes that are concentric with the axis of rotation. Unlike a turbine, which also employ rotor blades and stator vanes the flow path of an axial compressor decrease in cross-sectional area in the direction of flow. This reduces the volume of air as compression progresses from stage to stage.

After being delivered to the face of the compressor by the air inlet duct incoming air passes through the inlet guide vanes. Upon entering the first set of rotating blades, the air, which is flowing in a general axial direction is deflected in the direction of rotation. The air is arrested and turned as it is passed on to a set of stator vanes. Following that it is picked up by another set of rotating blades and soon through the compressor. Air pressure increases each time it passes through a set of rotors and stators.

The rotor blades increase the air velocity. When air velocity increases, the ram pressure of air passing through a rotor stage also increases. This increase in velocity and pressure is somewhat but not entirely nullified by diffusion. When air is forced past the thick sections of the rotor blades static pressure also increases. The larger area at the rear of the blades (due to its airfoil shape) acts as a diffuser.

In the stators velocity decreases while static pressure increases. As air velocity decreases, the pressures due to velocity or ram that has just been gained in passing through preceding rotor stage decreases somewhat; however, the total pressure is the sum of static pressure and pressure due to ram. Successive increases and decreases in velocity as air leaves the compressor are usually only slightly greater than the velocity of the air at the entrance to the compressor. As the pressure is built up by successive sets of rotors and stators, less and less volume is required. Thus, the volume within the compressor is gradually decreased. At the exit of the compressor, a diffuser section adds the final stage to the compression process by again decreasing velocity and increasing static pressure just before the air enters the engine burner section.

Normally, the temperature change caused by diffusion is not significant by itself. The temperature rises which causes air to get hotter and hotter as it continues through the compressor, is the result of the work being done on the air by the compressor rotors. Heating of the air occurs because of the compression process and because some of the mechanical energy of the rotor is converted to heat energy.

Because airflow in an axial compressor is generally diffusing it is very unstable. High efficiency is maintained only at very small rates of diffusion. Compared to a turbine, quite a number of compressor stages are needed to keep the diffusion rate small through each individual stage. Also, the permissible turning angles of the blades are considerably smaller than those which can be used in turbines. These are the reasons why an axial compressor must have many more stages than the turbine which drives it. In addition, more blades and consequently more stages are needed because the compressor, in contrast to a turbine, is endeavoring to push air in a direction than it does not want to go in.

Vocabulary

turbine	/ˈtɜːbaɪn/	n.	涡轮机;汽轮机;透平机
combustion	/kəmˈbʌstʃən/	n.	燃烧,烧毁;氧化;骚动
appropriate	/əˈprəʊpriət , əˈprəʊprieɪt/	adj.	适当的;合适的
		v.	私占;盗用;挪用;拨出(专款)
accessories	/əkˈsesəriz/	n.	附件;附件(accessory 的名词复数); (衣服的)配饰;从犯;妇女饰品
pressurization	/ˌpreʃəraɪˈzeɪʃn/	n.	压力输送;挤压;气密;增压
volume	/ˈvɒljuːm/	n.	体积;音量;一卷;合订本

		adj.（时间或地点上）在先的，在前的，前面的；前述的；上述的
preceding	/prɪˈsiːdɪŋ/	*v.* 在……之前发生（或出现）；走在……前面；先于；"precede"的现在分词
servo	/ˈsɜːvəʊ/	*n.* 伺服，伺服系统；继动
running instruments		运行仪器
the type of compressor		压缩机类型
multistage	/ˈmʌltɪsteɪdʒ/	*adj.* 多级的；多段；多工位
stator	/ˈsteɪtə/	*n.* 定子，固定片
rotor	/ˈrəʊtə(r)/	*n.* 轮子，旋转器；旋翼
dual-axial	/dˈuːəlˈæksɪəl/	*n.* 双轴
centrifugal	/ˌsentrɪˈfjuːgl/	*adj.* 离心的

┊ **Exercise** ┊

Complete the following sentences and translate them into Chinese.

（1）Bleed air has various uses including driving the remote-driven accessories. Some current applications of bleed air are:＿＿＿＿＿＿＿＿＿＿＿＿＿.

翻译：＿＿＿＿＿＿＿＿＿＿＿＿＿＿＿＿＿＿＿＿＿＿＿＿＿＿＿＿＿＿＿＿

＿＿＿＿＿＿＿＿＿＿＿＿＿＿＿＿＿＿＿＿＿＿＿＿＿＿＿＿＿＿＿＿＿＿＿＿

（2）Centrifugal-flow compressors have the following advantages:＿＿＿＿＿＿＿＿.

翻译：＿＿＿＿＿＿＿＿＿＿＿＿＿＿＿＿＿＿＿＿＿＿＿＿＿＿＿＿＿＿＿＿

＿＿＿＿＿＿＿＿＿＿＿＿＿＿＿＿＿＿＿＿＿＿＿＿＿＿＿＿＿＿＿＿＿＿＿＿

（3）Axial-flow compressor have two main elements:＿＿＿＿＿＿＿＿＿＿.

翻译：＿＿＿＿＿＿＿＿＿＿＿＿＿＿＿＿＿＿＿＿＿＿＿＿＿＿＿＿＿＿＿＿

＿＿＿＿＿＿＿＿＿＿＿＿＿＿＿＿＿＿＿＿＿＿＿＿＿＿＿＿＿＿＿＿＿＿＿＿

（4）The blades are attached in the disc rim by different methods using either the ＿＿＿＿＿＿ or bulb-type root designs.

翻译：＿＿＿＿＿＿＿＿＿＿＿＿＿＿＿＿＿＿＿＿＿＿＿＿＿＿＿＿＿＿＿＿

＿＿＿＿＿＿＿＿＿＿＿＿＿＿＿＿＿＿＿＿＿＿＿＿＿＿＿＿＿＿＿＿＿＿＿＿

（5）Axial-flow compressors have the following advantages:＿＿＿＿＿＿＿＿＿.

翻译：＿＿＿＿＿＿＿＿＿＿＿＿＿＿＿＿＿＿＿＿＿＿＿＿＿＿＿＿＿＿＿＿

＿＿＿＿＿＿＿＿＿＿＿＿＿＿＿＿＿＿＿＿＿＿＿＿＿＿＿＿＿＿＿＿＿＿＿＿

（6）Using the centrifugal-flow compressor boosts compression and＿＿＿＿＿＿.

翻译: _____

_____

(7) Normally, the temperature change caused by_____ is not significant by itself.

翻译: _____

_____

(8) In addition the blades must res is t bending loads_____.

翻译: _____

_____

+ Supplementary Reading +

Dual Compressor

The dual compressor is a combination either of two axial compressors or of an axial and a centrifugal compressor. The dual-axial compressor consists of a low-pressure compressor in front and a high-pressure compressor in the rear. Both compressor (low and high) are driven by two different shafts that connect to different turbines. The starter is usually connected to the high-pressure compressor because it reduces the torque required to start the engine. With the rear (high-pressure) compressor turning at governed speed, the front (low-pressure) compressor (not governed) is automatically rotated by its turbine. Rotation speed is whatever speed will ensure an optimum flow of air through the compressor. With the front and rear compressor rotors working in harmony instead of interfering with each other, compression rates can be increased without decreasing efficiency. Due to the added length of the engine this type of compressor is found on turbojet aircraft.

Most gas turbines in army aircraft have a combination of an axial compressor (front) and a centrifugal compressor (rear). The usual combination is a five-or seven-stage axial—flow compressor and a centrifugal-flow compressor. The axial compressor and centrifugal compressor combination is mounted on the same shaft; the compressors turn in the same direction and at the same speed. By combining them, the manufacturer makes the most of the advantages of both compressors small frontal area, increased compression ratios, and shortened overall engine length. Using the centrifugal-flow compressor boosts compression and increases efficiency of the turbine engine. The centrifugal compressor also shortens the length of the engine. If the centrifugal compressor were not added, the manufacturer would have to add more stages of axial compression to equal that of the centrifugal compressor.

Lesson 17　Combustion Chamber

The combustion section contains the combustion chambers, igniter plugs, and fuel nozzle or fuel injectors. It is designed to burn a fuel-air mixture and to deliver combusted gases to the turbine at a temperature not exceeding the allowable limit at the turbine inlet. Theoretically, the compressor delivers 100 percent of its air by volume to the combustion chamber. However, the fuel-air mixture has a ratio of 15 parts air to 1part fuel by weight. Approximately 25 percent of this air is used to attain the desired fuel-air ratio. The remaining 75 percent is used to form an air blanket around the burning gases and to dilute the temperature, which may reach as high as 3,500°F, by approximately one-half. This ensures that the turbine section will not be destroyed by excessive heat.

The air used for burning is known as primary air; that used for cording is secondary air. Secondary air is controlled and directed by holes and louvers in the combustion chamber liner. Igniter plugs function during starting only; they are shut off manually or automatically. Combustion is continuous and self-supporting. After engine shutdown or failure to start, a pressure-actuated valve automatically drains any remaining unburned fuel from the combustion chamber. The most common type used in Army gas turbine engines is the external annular reverse-flow type.

The primary function of the combustion section is, of course, to bum the fuel-air mixture, thereby adding heat energy to the air. To do this efficiently, the combustion chamber must:

- Provide the means for mixing the fuel and air to ensure good combustion;
- Bum this mixture efficiently;
- Cool the hot combustion products to a temperature which the turbine blades can withstand under operating conditions;
- Deliver the hot gases to the turbine section.

The location of the combustion section is directly between the compressor and turbine sections. The combustion chambers are always arranged coaxially with the compressor and turbine. Regardless of type, since the chambers must be in a through-flow position to function efficiently.

All combustion chambers contain the same basic elements:

- A casing;
- A perforated inner liner;

- A fuel injection system;
- Some means for initial ignition;
- A fuel drainage system to drain off unburned fuel after engine shutdown.

There are currently three basic types of combustion chambers, varying detail only:

- The multiple-chamber or can type;
- The annular or basket type;
- The can-annular type.

Can-Type Combustion Chamber

The can-type combustion chamber is typical of the type used on both centrifugal and axial-flow engines. It is particularly well suited for the centrifugal compressor engine since the air leaving the compressor is already divided into equal portions as it leaves the diffuser vanes. It is then a simple matter to duct the air from the diffuser into the respective combustion chambers arranged radially around the axis of the engine. The number of chambers will vary; in the past as few as 2 and as many as 16 chamber have been used . The present trend is about 8 or 10 combustion chambers. On American built engines these chambers are numbered in clockwise direction facing the rear of the engine with the No. 1 at the top.

Each can-type combustion chamber consists of an outer case or housing with a perforated stainless steel (highly heat-resistant) combustion chamber liner or inner liner. The outer case is divided for ease of line replacement. The larger section or chamber body encases the liner at the exit end; the smaller chamber cover encases the front or inlet end of the liner. The inter-connector (flame propagation) tubes are necessary part of can-type combustion chambers. Since each can is a separate burner operating independently of the others, there must be some way to spread combustion during the initial starting operation. This is done by inter-connecting all the chambers; it passes through the tubes and igniter plugs in two of the lower chambers; it passes through the tubes and ignites the combustible mixture in the adjacent chamber. This continues until all chambers are burning. The flame tubes will vary in construction details from one to another although the basic components are almost identical.

Bear in mind that not only must the chambers be interconnected by an outer tube (in this case, a ferrule), but there must also be a slightly longer tube inside the outer one to interconnect the chamber liners where the flame is located the outer tubes or jackets around the inter-connecting flame. Tubes not only afford airflow between the chambers but also fulfill an insulating function around the hot flame tubes.

The spark igniters are normally two in number. They are located in two of the can-type combustion chambers.

Another very important requirement in the construction of combustion chambers is providing the means for draining unburned fuel. This drainage prevents gum deposits in the fuel manifold, nozzles and combustion chambers. These deposits are caused by the residue left when fuel evaporates. If fuel is allowed to accumulate after shutdown there is the danger of after fire.

If the fuel is not drained, a great possibility exists that at the next starting attempt excess fuel in the combustion chamber will ignite and tailpipe temperature will go beyond safe operating limits.

The liners of can-type combustors have perforations of various sizes and shapes, each hole having a specific purpose and effect on flame propagation in the liner. Air entering the combustion chamber is divided by holes, louvers, and slots into two main streams—primary and secondary air. Primary (combustion) air is directed inside the liner at the front end where it mixes with the fuel and bums. Secondary (cooling) air passes between the outer casing and the liner and joins the combustion gases through larger holes toward the rear of the liner, cooling the combustion gases from about 3, 500°F to near 1, 500°F。

Holes around the fuel nozzle in the dome or inlet end of the can-type combustor liner aid in atomization of the fuel. Louvers are also provided along the axial length of the liners to direct a cooling layer of air along the inside wall of the liner. This layer of air also tends to control the flame pattern by keeping it centered in the liner, preventing burning of the liner walls.

Annular-or Basket-Type Combustion Chamber

Some provision is always made in the combustion chamber case or in the compressor air outlet elbow for installation of a fuel nozzle. The fuel nozzle delivers the fuel into the liner in a freely atomized spray. The freer the spray, the more rapid and efficient the burning process. Two types of fuel nozzles currently being used in the various types of combustion chambers are the simplex nozzle and the duplex nozzle.

The annular combustion chamber consists basically of a housing and a liner as does the can-type. The liner consists of an undivided circular shroud extending all the way around the outside of the turbine shaft housing. The chamber may be constructed of one or more baskets. If two or more chamber are used, one is placed outside the other baskets. If two or more chamber are used, one is placed outside the other in the same radial plane, the term "double-annularchamber".

The spark igniter plugs of the annular combustion chamber are the same basic type used in the can combustion chambers, although construction may vary. There are usually two plugs mounted on the boss provided on each of the chamber housings. The plugs must be long enough to protrude from the housing into the outer annulus of the double-annular combustion chamber.

The annular-type combustion chamber is used in many engines designed to use the axial-flow compressor. It is also used by engines incorporating dual-type compressors (combinations of axial flow and centrifugal flow). Its use permits building an engine of small diameter. Instead of individual combustion chambers, the compressed air is introduced into an annular space formed by a combustion chamber liner around the turbine shaft. Usually, enough space is left between the outer liner wall and the combustion chamber housing to permit the flow of cooling air from the compressor. Fuel is introduced through nozzles or injectors connected to a fuel manifold. The nozzle opening may face upstream or downstream to airflow depending on engine design. Various means are provided to introduce primary (compressed) air to the vicinity of the nozzle or injectors to support combustion and additional air downstream to increase the mass flow. The liner of this type of burner consists of continuous, circular, inner and outer shrouds around the outside of the compressor drive shaft housing. Holes in the shrouds allow secondary cooling air to enter the center of the combustion chamber. Fuel is introduced through a series of nozzles at the upstream end of the liner. Because of their proximity to the flames all types of burner liners are short-lived in comparison to other engine components; they require more frequent inspection and replacement.

This type of burner uses the limited space available most effectively, permitting better mixing of the fuel and air within a relatively simple structure. An optimum ratio of burner inner surface area to volume is provided; this ensures maximum cooling of the gases as combustion occurs the design also tends to prevent heat warping. However, the burner liner on some engines from cannot be disassembled without removing the engine the aircraft—a distinct disadvantage.

The latest annular combustion system for military use is a low-pressure fuel injection system with vortex air swirlers to mix fuel and compressor discharge air before combustion. The fuel injector is positioned into the center of an air swirler in the dome of the liner. Fuel leaving the injectors (which has been swirled) is surrounded by a concentric air vortex pattern this breaks fuel particles down to an extremely small size before they reach the combustion zone. This creates excellent fuel-air mixing that ensures a low smoke level in the exhaust. The low-pressure fuel system does hot have tine nozzle orifices and can handle contaminated fuel without clogging.

Can-Annular-Type Combustion Chamber

The can-annular-type combustion chamber was developed by Pratt and Whitney for use in their JT3 axial-flow turbojet engine. Since this engine features the split-spool compressor, if needed a combustion chamber capable of meeting the stringent requirements of maximum strength and limited length plus high overall efficiency. These were necessary because of the high air pressures and velocities in a split-spool compressor along with the shaft length limitations explained below.

The split compressor requires two concentric shafts to join the turbine stages to the their respective compressors. The front compressor joined to the rear turbine stages requires the longer shaft. Because this shaft is inside the other, a limitation is imposed on diameter. The distance between the front compressor and the rear turbine must be limited if critical shaft lengths are to be avoided.

Since the compressor and turbine are not susceptible to appreciable shortening the necessary shaft length limitation had to be absorbed by developing a new type of burner. A design was needed that would give desired performance in much less relative distance than had previously been assigned. Can-annular combustion chambers are arranged radially around the axis of the engine in this instance the rotor shaft housing. The combustion chambers are enclosed in a removable steel shroud that covers the entire burner section. This feature makes the burners readily available for any required maintenance. The burners are interconnected by projecting flame tubes. These tubes make the engine-starting process easier. They function identically with those previously discussed but differ in construction details.

Each combustion chamber contains a central bullet-shaped perforated liner. The size and shape of the holes are designed to admit the correct quantity of air at the correct velocity and angle. Cutouts are provided in two of the bottom chambers for installation of the spark igniters. The combustion chambers are supported at the aft end by outlet duct clamps. These clamps secure them to the turbine nozzle assembly.

The forward face of each chamber presents six apertures which align with the six fuel nozzles of the corresponding fuel nozzle duster. These nozzles are the dual-orifice (duplex) type. They require a flow divider (pressurizing valve) as was mentioned above in the can type combustion chamber discussion. Around each nozzle are preswirl vanes for imparting a swirling motion to the fuel spray. This results in better atomization burning and efficiency.

Swirl vanes perform two important functions. They cause:

• High flamespeed-provides better nixing of air and fuel and ensures spontaneous burning;

• Low air velocity axially-swirling prevents the flame from moving axially too rapidly.

Swirl vanes greatly aid flame propagation because a high degree of turbulence in the early combustion and cooling stage is desirable. Vigorous mechanical mixing of fuel vapor with primary air is necessary; mixing by diffusion alone is too slow. Mechanical mixing is also done by other measure; for example, placing coarse screens in the diffuser outlet as is done in most axial-flow engines.

Can-annular combustion chambers must also have fuel drain valves in two or more of the bottom chambers. This ensures drainage of residual fuel to prevent its being burned at the next start.

The flow of air through the holes and louvers of the can-annular chambers is almost identical with the flow through other types of burners. Special baffling is used to swirl the combustion airflow and to give it turbulence.

Vocabulary

igniter	/ɪɡˈnaɪtər/	n. 点火器
nozzle	/ˈnɒzl/	n. 管嘴，喷嘴
excessive-heat		过热
coaxially	/ˈkəʊksɪəlɪ/	adv. 同轴地
through-flow		直流，通流
compressor	/kəmˈpresə(r)/	n. 压气机，压缩机
inter-connector	/ɪntəkəˈnektə/	n. 中继馈（电）线，内部连线；内连线
louvers	/ˈluːvəz/	n. 百叶窗
protrude	/prəˈtruːd/	vt. 使突出；使伸出
		vi. 突出；伸出
apertures	/æˈpətʃəz/	n. 孔（aperture 的名词复数）；隙缝；（照相机的）光圈；孔径
aerodynamic drag		气动阻力
vortex air swirler		涡旋空气交换机
axial-flow engines		轴流式发动机
low-pressure fuel system		低压燃油系统

Exercise

Complete the following sentences and translate them into Chinese.

(1) The combustion section contains the combustion chambers_____

_____.

翻译：_____

_____

(2) The remaining 75 percent is used to form an air blanket around the burning gases and to dilute the temperature，which may reach as high as _____，by approximately one-half. This ensures that the turbine section will not be destroyed by_____.

翻译：_____

_____

(3) The location of the combustion section is directly between_____ and

_____.

翻译：_____

_____

(4) The split compressor requires two concentric shafts to join the turbine stages to the

_____.

翻译：_____

_____

(5) Since the compressor and turbine are not susceptible to appreciable shortening the necessary shaft length limitation had to_____.

翻译：_____

_____

(6) The forward face of each chamber presents six apertures which align with_____

_____.

翻译：_____

_____

(7) Can-annular combustion chambers must also have_____ in two or more of the bottom chambers.

翻译：_____

_____

Supplementary Reading

Performance Requirements of Combustion Chamber

Performance requirements include the following：

- High combustion efficiency. This is necessary for long range.

- Stable operation. Combustion must be free form blowout at airflows ranging from idle to maximum power and at pressures representing the aircraft's entire altitude range.

- Low pressure loss. It is desirable to have as much pressure as possible available in the exhaust nozzle to accelerate the gas rearward high pressure losses will reduce thrust and in-crease specific fuel consumption.

• Uniformtemperature dist ribution.

• The average temperature of gases entering the turbine should be as close as possible to the temperature limit of the burner material to obtain maximum engine performance. High local temperatures or hot sports the gas stream will reduce the allowable average turbine inlet temperature to protect the turbine. This will result in a decrease in total gas energy and a corresponding decrease engine performance.

• Easy staring. Low pressures and high velocities in the burner make starting difficult. A poorly designed burner will start only within a small range of light speeds and altitudes, whereas a well designed burner will permit easier air restarts.

• Small size. A large burner requires a large engine housing with a corresponding increase in the airplane's frontal area and aerodynamic drag. This will result in a decrease in maximum flight speed excessive burner size also results in high engine weight, lower fuel capacity and payload, and shorter range. Modem burners release 500 to 1, 000 times the heat of a domestic oil burner or heavy industrial furnace of equal unit volume. Without this high heat release the aircraft gas turbine could not have been made practical.

• Low-smoke burner. Smoke not only annoys people on the ground. It may also allow easy tracking of high-flying military aircraft.

• Low carbon formation. Carbon deposits can block critical air passages and disrupt airflow along the liner walls, causing high metal temperatures and low burner life.

All of the burner requirements must be satisfied over a wide range of operating conditions. For example, airflows may vary as much as 50 : 1, fuel flows as much as 30 : 1, and fuel-air ratios as much as 5 : 1. Burner pressures may cover a ratio of 100 : 1, while burner inlet temperatures may vary by more than 700° F. The effects of operating variables on burner performance are:

• pressure;

• Inletair temperature;

• Fuel-air ratio;

• Flow velocit.

Lesson 18　Turbine

All turbines in modern jet engines, regardless of the type of compress used, are of axial-flow design. They consist of one or more stages located immediately to the rear of the engine burner section. Turbines extract kinetic energy from the expanding gases as the gases come from the burners. They convert this energy in to shaft horsepower to drive the compressor and engine accessories. In a turboshaft or turboprop engine one or more turbines will also furnish the power required to turn the engine drive or propeller shaft. Nearly three-fourths of all of the energy available from combustion is needed to drive the compressor or compressor in the case of a dual-compressor engine. This includes the fan of a turbofan engine. If the engine is a turboshaft or turboprop, the turbines are designed to extract as much energy as possible from the gases passing through the engine. So efficient are the turbines in such engines that the propeller in a turboprop aircraft provides approximately 90 percent of the propulsive force with only 10 percent supplied by jet thrust. The axial-flow turbine has two main elements; turbine rotors (or wheels, as they are sometimes called) and stationary vanes. The stationary part of the assembly consists of a plane of contoured vanes, concentric with the axis of the turbine and set at an angle to form a series of small nozzles. These nozzles discharge the gases onto the blades in the turbine rotors. The stationary vane assembly of each stage in the turbine is usually referred to as the turbine nozzle guide vanes. The turbine nozzle area is the most critical part of the turbine design. If the nozzle area is too large, the turbine will not operate at its best efficiency. If the area is too small the nozzle will have a tendency to choke and lose efficiency under maximum thrust conditions. The turbine nozzle area is defined as the total cross-sectional area of the exhaust gas passages at their narrowest point through the turbine nozzle. It is calculated by measuring and adding the areas between individual nozzle guide vanes.

There are three types of turbines: impulse reaction and a combination of these two, known as reaction-impulse. In the impulse type there is no net change in pressure between the rotor inlet and the rotor exit. The blade's relative discharge velocity will be the same as its relative inlet velocity. The nozzle guide vanes are shaped to form passages which increase the velocity and reduce the pressure of the escaping gases. In the reaction type, the nozzle guide vanes do little more in relation to the rotor than alter flow direction. The decrease in pressure and increase in velocity of gases are caused by the convergent shape of the passage between the rotor blades. In a jet engine the turbine is usually a balanced combination of both types known as a reaction-impulse turbine. Its design is intended to achieve both a small diameter and a proper match with the compressor. Turbines may be either single or multiple stages. When the turbine has more than one stage, stationary vanes are inserted between each rotor wheel and the rotor wheel downstream. They are also placed at the entrance and exit of the turbine unit. Each set of stationary vanes forms a nozzle-vane assembly for the turbine wheel that follows. The exit set of vanes servers to straighten the gas flow before passage through the jet nozzle. The wheels may or may not operate independently of one another, depending on engine type and turbine power requirements.

Shaft RPM, gas flow rate, turbine inlet and outlet temperature and pressure turbine exhaust velocity, and required power output must all be considered by the designer of the turbine. If the engine is equipped with a dual compressor, the turbine must also be dual or "split". In this event, the forward part of the turbine (which drives the high-pressure compressor) can be single-stage because it receives high-energy gases directly form the burner and turns at a higher RPM than the turbine for the low-pressure compressor. By the time the gases reach the rear part of the turbine (which drives the low-pressure compressor), they have expanded. Considerably more blade area is needed if work or energy balance is to be maintained. To do a multistage turbine is used for the second part of the turbine (Figure 18. 1).

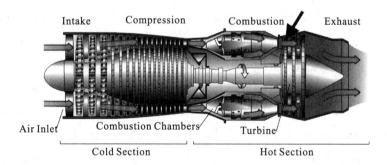

Figure 18. 1   Dual-Rotor Turbine for Split-Spool Compressor

Turbines must be designed so that the gases have a high expansion ratio. This results in a large temperature drop in gases passing through the turbine and a cool turbine exhaust. If the engine is equipped with an afterburner, a cool exhaust enables more fuel to be burned in the afterburner without exceeding the temperature limit, of the construction materials used in the afterburner.

The turbine wheel is a dynamically balanced unit consisting of super alloy blades attached to a rotating disc. The base of the blade is usually a "fir tree" design to enable it to be firmly attached to the disc and still allow room for expansion. In some turbines the rotating blades are open at their outer perimeter. More commonly, the blade is shrouded at the tip. The shrouded blades form a band around the perimeter of the turbine wheel, which serves to reduce blade vibrations. The weight of the shrouded tips is offset because the shrouds permit thinner, more efficient blade sections than are otherwise possible because of vibration limitations. Also, by acting in the same manner as aircraft wing tip fences, the shrouds improve the airflow characteristics and increase the efficiency of the turbine. The shrouds also serve to cut down gas leakage around the tips of the turbine blades. Turbines are subjected to high speeds and high temperatures. High speeds result in high centrifugal forces. Turbines must operate close to temperature limits that, if exceeded, lower the strength of the materials they are constructed of Turbine blades undergo distortion or lengthening known as "creep". Creep means that the blade stretches or elongates. This condition is cumulative. The rate of creep is determined by the load imposed on the turbine and the strength of the blade. The strength of the blade is determined by the temperature within the turbine. Since changes in pitch and creep are more pronounced if engine operating limits are not respected the pilot or flight engineer must closely observe the temperature and RPM limits stipulated by the manufacturer.

Construction

The turbine wheel is one of the most highly stressed engine parts. Not only must it operate at temperatures up to approximately $1,700°F$, but it must do so under severe centrifugal load s imposed by high rotational speeds of over $40,000$ RPM for small engines and $8,000$ RPM for larger ones. Consequently, engine speed and turbine inlet temperature must be accurately controlled to keep the turbine within safe operating limits. The turbine assembly is made of two main parts: the disc and blades. The disc or wheel is a statically and dynamically balanced unit of specially alloyed steel. It usually contains large percentages of chromium, nickel and cobalt. After forging, the disc is machined all over and carefully inspected using X rays magnetism, and other inspection methods for structural integrity. The blades or buckets are attached to the disc by means of a fir tree design go allow for different rates of expansion between the disc and the blade while still holding the blade firmly against centrifugal loads. The blade is kept from moving axially either by rivets, special locking tabs or devices, or another turbine stage.

Some turbine blades are open at the outer perimeter; in other a shroud is used. The shroud acts to prevent blade-tip losses (gas leakage around the tips of the turbine blade) and excessive vibration. By acting in the same manner as aircraft wing tip fence, the shrouds improve airflow characteristics and increase turbine efficiency (Figure18. 2).

Figure18. 2   Turbine Blade Tips

Shrouds reduce resistance to distortion under high loads, which tend to twist the blade toward low pitch. The shrouded blade has an aerodynamic advantage; thinner blade sections can be used and tip lines can be reduced by using a knife-edge or labyrinth seal at this point.

Shrouding however, requires that the turbine run cooler or at reduced RPM because of the extra mass at the tip. On blades that are not shrouded, the tips are cut or recessed to a knife-edge to permit a rapid "wearing-in" of the blade tip to the turbine casing with am-responding increase in turbine efficiency.

Blades are forged from highly alloyed steel. They are carefully machined and inspected before being certified for use. Many engine manufacturers will stamp a moment weight number on the blade to retain rotor balance when replacement is necessary.

Another method for increasing efficiency is the use of honeycomb shrouding. This shroud works as a labyrinth sealing the unshrouded turbine tips. These shrouds are all housed by a stator support, which, in turn, is supported by the engine outer casing. This design is currently in use in the new General Electric turboshaft engines.

Nozzle vanes may be either case or forged. Some vanes are hollow to allow a degree of cooling by compressor bleed air. In all cases the nozzle assembly is made of very high-temperature, high-strength, steel to withstand the direct impact of the hot high-pressure, high-velocity gas flowing from the combustion chamber (Figure 18. 3).

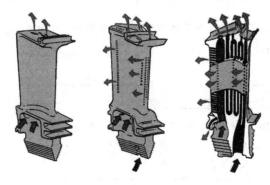

Figure 18. 3   Turbine Nozzle Cooling

Some manufacturers are experimenting with the engine with transpiration-cooled nozzle and turbine blading in which the airflows through thousands of small holes in a porous airfoil made from a sintered wire mesh material. The performance of the gas turbine engine depends largely on the temperature at the inlet. Increasing this temperature from the present limit of about 1,750°F to the 2,500°F possible with transpiration-cooled blades will result in about a 100 percent increase in specific horsepower. Transpiration cooling may be a promising development in gas turbine design.

Vocabulary

turbines	/ˈtɜrˌbaɪn/	n. 涡轮机
turboshaft	/ˈtɜrboʊˌʃæft/	n. (驱动水泵等的)涡轮轴发动机
turboprop	/ˈtɜːbəʊprɒp/	n. 涡轮螺旋桨发动机
velocity	/vəˈlasəti/	n. 速率，速度
afterburner	/ˈɑːftəbɜːnə(r)/	n. 喷射引擎等的加力燃烧室；补燃器；后燃室
efficiency	/ɪˈfɪʃ(ə)nsi/	n. 效率；功率
manufacturers	/ˌmænjəˈfæktʃərər/	n. 制造商，制造厂（manufacturer 的名词复数）
transpiration-cooled		蒸发冷却的，蒸发冷却式
distributed	/dɪsˈtrɪbjuːtɪd/	adj. 分布式的
shrouded	/ʃraʊd/	v. 隐瞒
operating	/ˈɒpəreɪtɪŋ/	adj. 操作的；营运的
shrouds	/ʃˈraʊdz/	n. 遮蔽物；覆盖物

Exercise

Complete the following sentences and translate them into Chinese.

(1) Another method for _____ the use of honeycomb shrouding.

翻译：_____

_____

(2) The stationary vane assembly of each stage in the _____ to as the turbine nozzle guide vanes.

翻译：_____

_____

(3) In a jet engine the turbine is usually a _____ of both types known as a reaction-impulse turbine.

翻译：_____

_____

(4) This results in a large temperature drop in _____ the turbine and a cool turbine exhaust.

翻译：_____

_____

(5) The turbine _____ alloy blades attached to a rotating disc.

翻译：_____

_____

(6) The shrouded blades form a band around the _____, which serves to reduce blade vibrations.

翻译：_____

_____

(7) This disc or wheel is a _____ of specially alloyed steel.

翻译：_____

_____

(8) _____. Although bleed air _____ be hot, it is cool in relation to the temperature at the turbine inlet. This air, therefore serves to cool the vanes and blades.

翻译：_____

_____

Supplementary Reading

Cooling

Design engineers use every device at their command to increase the allowable inlet temperature. On practically all large engines, one such device is to cool the fret-stage turbine inlet guide vanes and the first-stage rotor blades. This is done by conducting compressor bleed air through passages inside the engine to the turbine area. There, the air (the coolant) is led to the longitudinal holes, tubes or cavities in the first-stage vanes and blades.

After entering the vane and blade passages, the air (coolant) is distributed through holes at the leading and trailing edges of the vanes and blades. The air impinges along the vane and blade surfaces and then passes out of the engine with the exhaust. Although bleed air coming from the compressor may be hot, it is cool in relation to the temperature at the turbine inlet. This air, therefore serves to cool the vanes and blades. These permits gases coming from the burner section to enter the turbine at higher temperatures than would otherwise be permissible.

Cooling is necessary only in the turbine inlet area because enough energy is extracted from the exhaust gases by the first or first and second stages of the turbine to reduce the temperature to a tolerable level.

Lesson 19   Exhaust

The term "exhaust duet" applies to the engine exhaust pipe or tail pipe including the jet nozzle of a non-after-burning engine (Figure 19.1). Although an afterburner might also be considered a type of exhaust duct, after burning is a subject in itself.

Figure 19.1   Exhaust Duct

If the engine exhaust gases could be discharged directly to the outside air in an exact axial direction at the turbine exit, an exhaust duct might not be necessary. This, however, is not practical. A larger total thrust can be obtained from the engine if the gases are discharged from the aircraft at a higher velocity than that permissible at the turbine outlet. An exhaust duct is added to collect and straighten the gas flow as it comes from the turbine. It also increases the velocity of the gases before they are discharged from the exhaust nozzle at the rear of the duct. Increasing gas velocity increases its momentum and the thrust produced.

An engine exhaust duct is often referred to as the engine tail pipe. The duct is essentially a simple, stainless steel, conical or cylindrical pipe. The engine tail cone and struts are usually included at the rear of the turbine. The struts support the rear bearing and impart an axial direction to the gas flow, the tail cone helps smooth the flow. Immediately after the turbine outlet and usually just forward of the flange to which the exhaust duct is attached, the engine has a sensor for turbine discharge pressure. In large engines, it is not practical to measure internal temperature at the turbine inlet. Therefore, the engine is usually also instrumented for exhaust gas temperature at the turbine outlet. One or more thermocouples preinserted in the exhaust case to provide adequate sampling of exhaust gases. Pressure probes are also inserted in this case to measure pressure of gases coming from the turbine. The gradually diminishing cross-sectional area of a conventional convergent type of exhaust duct capable of keeping the flow through the duct constant at velocities not exceeding Mach 1. 0 at the exhaust nozzle.

Exhaust Ducts

Turboshaft engines in helicopters do not develop thrust using the exhaust duct. If thrust were developed by the engine exhaust gas, it would be impossible to maintain a stationary hover; therefore, helicopters use divergent ducts. These ducts reduce gas velocity and dissipate any thrust remaining in the exhaust gases. On find-wing aircraft, the exhaust duct may be the convergent type, which accelerates the remaining gases to produce thrust. This adds additional SHP to the engine rating. Equivalent shaft horsepower (ESHP) is the combination of thrust and SHP.

Conventional Convergent Exhaust Nozzle

The rear opening of the exhaust duct is the jet nozzle, or exhaust nozzle as it is often called The nozzle acts as an orifice, the size of which determines velocity of gases as they emerge from the engine. In most non-after-burning engines, this area is critical; for this reason, it is fixed at the time of manufacture. The exhaust (jet) nozzle area should not be altered in the field because any change in the area will change both the engine performance and the exhaust gas temperature. Some early engines however, were trimmed to their correct RPM or exhaust gas temperature by altering the exhaust-nozzle area. When this is done, small tabs that may be bent as required are provided on the exhaust duct at the nozzle opening. Or small, adjustable pieces called "mice" are fastened as needed around the perimeter of the nozzle to change the area. Occasionally, engines are equipped with variable area nozzle which are opened or closed, usually automatically, with an increase or decrease in fuel flow. The velocity of the gases within a convergent exhaust duct is usually held to a subsonic speed. The velocity at the nozzle approaches Mach 1. 0 (the velocity at which the nozzle will choke) on turbojets and low-bypass-ratio turbofans during most operating conditions.

Convergent-Divergent Exhaust Nozzle

Whenever the pressure ratio across an exhaust nozzle is high enough to produce gas velocities which might exceed Mach 1. 0 at the engine exhaust nozzle, more thrust can be gained by using a convergent-divergent type of nozzle. This can be done provided the weight penalty is not so great that the benefit of the additional thrust is nullified. The advantage of a convergent-divergent nozzle (C-D nozzle) is greatest at high Mach numbers because of the resulting higher pressure ratio across the engine nozzle. If the pressure ratio through a subsonic exhaust duct is great enough (this will be the case when the pressure at the entrance to the exhaust duct becomes approximately twice that at the exhaust nozzle），the change in velocity through the duct will be enough to cause sonic velocity（Mach 1. 0）at the nozzle. At very high flight Mach numbers，the pressure ratio becomes much more than 20. If a C-D nozzle is used，the velocity at the exhaust nozzle becomes correspondingly greater than Mach 1. 0. This is a distinct advantage，provided the nozzle can effectively handle these high velocities.

When a divergent duct is employed in combination with a conventional exhaust duct，it is called a convergent-divergent exhaust duct. In the C-D nozzle，the convergent section is designed to handle the gases while they remain subsonic and to deliver them to the throat of the nozzle just as they attain sonic velocity. The divergent section handles the gases after they emerge from the throat and become supersonic further increasing their velocity.

Vocabulary

exhaust	/ɪɡˈzɔːst/	v. 耗尽；使精疲力竭；详尽地讨论
		n. 废气；排气管
nozzle	/ˈnɒzl/	n. 管嘴，喷嘴；〈俚〉鼻子
afterburner	/ˈɑːftəbɜːnə(r)/	n. 喷射引擎等的加力燃烧室；补燃器；后燃室
adequate	/ˈædɪkwət/	adj. 足够的；合格的；合乎需要的
turboshaft	/ˈtɜːbəʊʃæft/	n.（驱动水泵等的）涡轮轴发动机
manufacture	/ˌmænjuˈfæktʃə(r)/	v. 制造，生产；捏造，虚构；加工；从事制造
		n. 大量制造；批量生产；工业品
convergent-divergent		收缩膨胀，缩放
correspondingly	/ˌkɒrəˈspɒndɪŋli/	adv. 相对地，比照地
aircraft	/ˈeəkrɑːft/	n. 飞机；航空器
manufacturer	/ˌmænjuˈfæktʃərə(r)/	n. 制造商，制造厂；厂主；[经]厂商
turboshaft	/ˈtɜːbəʊʃæft/	n.（驱动水泵等的）涡轮轴发动机

Exercise

Complete the following sentences and translate them into Chinese.

（1）The term "_____" applies to the engine exhaust pipe or tail pipe including the jet nozzle of a non-after-burning engine.

翻译：_____

_____

(2) A larger total thrust can be obtained from the engine if the gases_____from the aircraft at_____than that permissible at the turbine outlet.

翻译：_____

_____

(3) Immediately aft the turbine outlet and usually just forward of the flange to which the _____, the engine has a sensor for turbine_____.

翻译：_____

_____

(4) If thrust were developed by the_____, it would be impossible to maintain a stationary hover; therefore, helicopters use_____.

翻译：_____

_____

(5) The rear opening of the exhaust duct is the jet nozzle, or_____as it is often called The nozzle acts as an orifice, the size of which_____as they emerge from the engine.

翻译：_____

_____

(6) The_____results in additional thrust which, as has been shown, must be added when the total thrust developed by the engine is computed.

翻译：_____

_____

(7) However, if the rate of change in the duct area is either too gradual or too rapid for the calculated increase in_____, unsteady flow_____of the throat will occur with an accompanying loss of energy.

翻译：_____

_____

(8) The solution to this dilemma is a_____with a variable_____configuration which can adjust itself to changing pressure conditions.

翻译：_____

_____

(9) As the actual design and operation of such nozzles is usually either classified _____ or _____ of the manufacturer, the nozzles cannot be

described here.

翻译:

Pressure generated within an engine cannot be converted to velocity, particularly when a convergent nozzle is used. The additional pressure results in additional thrust which, as has been shown, must be added when the total thrust developed by the engine is computed. The additional thrust is developed inefficiently. It would be much better to convert all of the pressure within the engine to velocity and develop all of the engine thrust by mean of changes in momentum. In theory, a C-D nozzle does this. Because it develops this additional part of the total thrust more efficiently, it enables an engine to produce more total net thrust than the same basic engine would generate if it were equipped with a conventional convergent duct and nozzle. The C-D nozzle would be nearly ideal if it could always be operated under the exact conditions for which it was designed. However, if the rate of change in the duct area is either too gradual or too rapid for the calculated increase in weight of the gases, unsteady flow downstream of the throat will occur with an accompanying loss of energy. This ultimately means loss of thrust. If the rate of increase in area of the duct is too little, the maximum gas velocity that can be reached will be limited. If the rate of increase is too great, the gas flow will break away from the surface of the nozzle, and the desired increase in velocity will not be obtained. As exhaust gases accelerate or decelerate with changing engine and flight conditions, their pressure fluctuates above or below the pressure ratio for which the nozzle was designed. When this occurs, the nozzle no longer converts all of the pressure to velocity, and the nozzle begins to lose efficiency.

The solution to this dilemma is a C-D nozzle with a variable cross-sectional configuration which can adjust itself to changing pressure conditions. Several types of C-D nozzles have been tried, and a few have been used successfully on production aircraft. As the actual design and operation of such nozzles is usually either classified military information or proprietary information of the manufacturer, the nozzles cannot be described here.

Lesson 20　　Thrust Reversers

The difficult problem of stopping an aircraft after landing increases many times with the greater gross weights common to large, modern aircraft with their higher wind loadings and increased landing speeds. Wheel brakes alone are no longer the beat way to slow the aircraft immediately after touchdown. The reversible-pitch propeller solved the problem for piston engine and turboprop-powered airplanes. Turbojet and turbofan aircraft, however, must rely on some device such as a parabrake or runway arrester gear or some means of reversing the thrust produced by their engines.

Although sometimes used on military aircraft, the parabrake or drag parachute has distinct disadvantage. The parabrake is always subject to either a premature opening or a failure to open at all. The parabrake must be recovered and repacked after each use and, if damaged or lost must be repaired or replaced. Once the parabrake has opened, the pilot has no control over the amount of drag on the aircraft except to release the parachute completely.

Arrester gears are primarily for aircraft carrier deck operation although they are sometimes used by military bases as overshoot barriers for land runways. They would hardly be suitable for commercial airline operation at a busy municipal airport.

Oil System Seals

The significance of oil system seals in aircraft engines is great. A leaking seal in a turbine engine could cause tire, bearing failure, or cockpit fumes, to name a few dangers. There are three main types of oil system seals: synthetic, labyrinth and carbon.

Synthetic

Synthetic seals (neoprene, silicone, teflon and synthetic rubber) are used throughout the engine1's oil system. They are used where metal-to-metal contact would not provide satisfactory sealing to withstand pressures in such items as filters, turbine and fittings. Seals come in many sizes and shapes and are not normally reused. New replacement seals are received from supply channels usually in a package that prevents damage. In most cases the packages will have a "cure date" stamped on the outside (Cure date is the date of manufacture of the seals.). This date is particularly important when installing seals retie of rubber, which has a tendency to deteriorate more rapidly than synthetic material. Just as important is to use the proper seal with the correct part number for a specific installation. Never use a seal from another system just because it looks like the right seal.

The composition or military specifications may be entirely different which could cause the seal to fail at a crucial moment.

Some synthetic seals coming into contact with synthetic oils such as MIL – 7808 or ML – 23609 have a tendency to swell; others might deteriorate completely. Occasionally, seals are referred to as "packings" or "gaskets". However, there is a difference between the two. Packing is used to provide a running seal; a gasket is used between two stationary parts to create a static seal. Some manufacturers refer to a gasket as a packing and vise versa. These terms should not be taken literally. Always go strictly by the part number when using seals.

Labyrinth

Labyrinth or air seals are designed to allow a small amount of air to flow across the sealing surface. This helps prevent oil (or lower – pressured air) seepage across the same surface. Air seals have two separate parts. One part forms a plain or honeycomb surface; the corresponding part is a circular seal with annular grooves. These grooves may use a soft metal as the basic composition or be machined into a surface. Matching the two together (one rotating portion or race with one stationary) forms an air pressure seal. A series of soft metal knife-like edges rides very close to the seal surface or cuts a path into a stationary honey comb or silver alloy air seal.

NOTE: When honeycomb or silver alloy is used, it is bonded to the stationary portion of the air seal.

Air for this seal is normally bled form the compressor and then forced between the sealing surface and seal. The effect of pressurization prevents oil (or lower-pressured air) from seeping from one section to another during engine operation. Air seals work only when the engine is operating when the engine is shut down seal leakage will occur. Be extremely careful when working in or around the seal area because seals are composed of very soft metal. Any small nick or groove in a seal may cause a serious oil leak which may require a premature engine change.

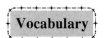
Vocabulary

loadings	/ˈləʊdɪŋz/	n. 负载；载荷；载分析；荷量；负荷量
parabrake	/ˈpærəbreɪk/	n. 减速(降落)伞，制动降落伞
arrester	/əˈrɛstə/	n. 逮捕者；避雷器；制动装置
turboprop-powered airplanes		涡轮螺旋桨飞机
significance	/sɪgˈnɪfɪkəns/	n. 意义；(尤指对将来有影响的)重要性；意思；含义

replacement	/rɪˈpleɪsmənt/	n. 替换；更换；替代品；替换物；（尤指工作中的）接替者；替代者
specification	/ˌspɛsɪfɪˈkeɪʃən/	n. 规格；规范；明细单；说明书
deteriorate	/dɪˈtɪəriəreɪt/	vi. 变坏；恶化；退化
manufacturer	/ˌmænjʊˈfæktʃərə/	n. 生产者；制造者；生产商
groove	/gruːv/	n. 沟；槽；辙；纹；（某种）音乐节奏 v. 在……上开出沟（或槽等）
centrifugal force		n. 离心力
configuration	/kənˌfɪgjʊˈreɪʃən/	n. 布局；结构；构造；格局；形状；（计算机的）配置
lower-pressured		低压
pressurization	/ˌprɛʃəraɪˈzeɪʃn/	n. 加压控制；加压；增压；加压送风；耐压
faceplate	/ˈfeɪsˌpleɪt/	n. 面板；花盘；（电灯开关、轴颈箱等的）保护罩；荧光屏；屏幕
seal race		密封圈

Exercise

Complete the following sentences and translate them into Chinese.

(1) Turbojet and turbofan aircraft, however, must rely on some device such as a _____ or _____ arrester gear or some means of reversing the thrust produced by their engines.

翻译：_____
_____

(2) The parabrake must be recovered and repacked after each use and, _____.

翻译：_____
_____

(3) Just as important is to use the proper seal with the correct part number for _____.

翻译：_____
_____

(4) One part forms a plain or honeycomb surface; the corresponding part is a circular seal with _____.

翻译：_____

(5) The steel surface is called a "_____" or "_____", depending on the engine manufacturer.

翻译：_____

(6) Three common preload methods are spring tension, centrifugal force, and _____.

翻译：_____

(7) Seals can be rebuilt by replacing the segments in groups; other seals, however, must be returned to an overhaul facility when _____.

翻译：_____

(8) These springs not only_____ but also serve as the preload necessary to press the seals inward.

翻译：_____

(9) On this particular seal, the seal race contact surface is _____ a spray of oil.

翻译：_____

(10) They would hardly be suitable for commercial airline operation at a _____.

翻译：_____

+·+·+·+·+·+·+·+·+·+·+·+·+·+·+
Supplementary Reading
+·+·+·+·+·+·+·+·+·+·+·+·+·+·+

Carbon

Carbon oil seals are used to contain the oil within the bearing areas on most jet turbine engines. All carbon seals form a sealing surface by having a smooth carbon surface rub against a smooth steel surface. The steel surface is called a "seal race" or "faceplate", depending on the engine manufacturer. All carbon seals are preloaded. That is, the carbon must in some way be pressed against the steel surface. Three common preload methods are spring tension, centrifugal force, and air pressure. During operation, the seal may be aided by allowing a small amount of oil to flow into the rubbing surface. The oil also cools the seal as a certain amount of heat is built up by the carbon rubbing on the steel surface.

The carbon oil seal consists of two rows of carbon segments (seal ring and back ring) mounted in a housing and held together around their circumference by extension springs.

These springs not only hold the segments together by circling the outside but also serve as the preload necessary to press the seals inward. The seal segments nearest the bearing have a lip that forms the seal; the positioning pads contact the steel race and maintain the proper sealing positions of the segments. These positioning pads are sometimes referred to as "wear blocks" because the seal oil is very thin and without the pads would have a short wear life. The grooves between the pads are staggered to reduce airflow toward the sump. On this particular seal, the seal race contact surface is cooled by a spray of oil. The key and spiral pin keep the carbon segments from turning within the seal housing the compression springs press the seal segments into the housing. The entire assembly is held together by the spring retainer and snap ring. This type of seal is stationary and rubs against an inner, rotating seal race.

Other configurations of carbon seals may have several seals on each side of the bearing. They may also rub on the side or outer surface of the seal rather than the inner surface as the one illustrated does. Seals can be rebuilt by replacing the segments in groups; other seals however, must be returned to an overhaul facility when they are damaged and must be replaced with a complete seal assembly.

Lesson 21   Materials

High-temperature, high-strength materials and unique methods of manufacture have made the gas turbine engine a practical reality in a few decades. The performance of turbojet and turboprop engines depends largely on the temperature at the inlet to the turbine. Increasing the turbine inlet temperature from the present limit (for most highly produced engines) of approximately 1,700°F to 2,500°F will result in a specific thrust increase of approximately 130 percent along with a corresponding decrease in specific fuel consumption. For this reason high cycle temperatures are desirable. Not all materials can withstand the hostile operating conditions found in parts of the gas turbine engine.

1. Properties and Characteristics

Metallurgists have been working for almost 50 years improving metals for use in aircraft construction. Each type of metal or alloy has certain properties and characteristics which make it desirable for a particular use, but it may have other qualities which are undesirable. The metallurgist's job is to build up the desirable qualities and tone down the undesirable ones. This is done by the alloying (combining) of metals and by various heat-treating processes. It is not necessary for the airframe repairer to be a metallurgist, but it is advantageous to have a general knowledge of the properties used in their development. The repairer should be familiar with a few metallurgical terms. The following terms are used in describing the physical properties and characteristics of metals.

 • Hardness. Hardness refers to the ability of a metal to resist abrasion, penetration, cutting action or permanent distortion. Hardness may be increased by working the metal and, in the case of steel and certain aluminum alloys, by heat-treatment and cold-working. Structural parts are often formed from metals in their soft state and are then heat-treated to harden them so the finished shape is retained. Hardness and strength are closely associated properties of metals.

 • Brittleness. Brittleness is the property of a metal which allows little bending or deformation without shattering. A brittle metal is apt to break or crack without change of shape. Because structural metals are often subjected to shock loads, brittleness is not a desirable property. Cast iron or cast aluminum and very hard steel are brittle metals.

 • Malleability. A metal which can be hammered, rolled or pressed into various shapes without cracking breaking or other detrimental effects is said to be malleable. This property is necessary in sheet metal which is worked into curved shapes such as cowling and wingtips. Copper is an example of a malleable metal.

• Ductility. Ductility is the property of a metal which permits it to be permanently drawn, bent or twisted into various shapes without breaking. This property is essential for metals used in making wire and tubing. Ductile metals are preferred for aircraft use because of their ease of forming and resistance to failure under shock loads. For this reason, aluminum alloys are used for cowl rings, fuselage and wing skin, and formed or extruded parts such as ribs, spars and bulkheads. Chrome-molybdenum steel is also easily formed into desired shapes. Ductility is similar to malleability.

• Elasticity. Elasticity is that property which enables a metal to return to its original shape when the force which causes the change of shape is removed. This property is extremely valuable, as it would not be desirable to have a part permanently distorted after an applied load was removed. Each metal has a point known as the elastic limit beyond which it can not be loaded without causing permanent distortion. In aircraft construction, members and parts are so designed that the maximum loads to which they are subjected will never stress them beyond their elastic limits. This desirable property is present in spring steel.

• Toughness. A material which possesses toughness will with stand tearing or shearing and may be stretched or otherwise deformed without breaking. Toughness is a desirable property in aircraft materials.

• Conductivity. Conductivity is the property which enables a metal to carry heat or electricity. The heat conductivity of a metal is especially important in welding as it governs the amount of heat that will be required for proper fusion. Conductivity of the metal, to a certain extent, determines the type of jig to be used to control expansion and contraction. In aircraft, electrical conductivity must also be considered in conjunction with bonding to eliminate radio interference. Metals vary in their capacity to conduct heat. Copper, for instance, has a relatively high rate of heat conductivity and is a good electrical conductor.

• Fusibility. Fusibility is the ability of a metal to become liquid by the application of heat. Metals are fused in welding. Steels fuse at approximately $2,500°F$ ($1,371℃$), and aluminum alloys fuse at approximately $1,100°F$ ($593℃$).

• Density. Density is the weight of a unit volume of a material. In aircraft work, the actual weight of a material per cubic inch is preferred. Density is an important consideration when choosing a material to be used in the design of a part in order to maintain the proper weight and balance of the aircraft.

• Contraction and expansion. Contraction and expansion are reactions produced in metals as the result of heating or cooling. A high degree of heat applied to a metal will cause it to expand or become larger. Cooling hot metal will shrink or contract it. Contraction and expansion affect the design of welding jigs, castings, and tolerances necessary for hot rolled material.

• Strength. Strength refers to the ability of a metal to hole loads (or forces) without breaking. Strength is a property that sums up many of the desirable qualities of metals. Strength with toughness is the most important combination of properties a metal can possess. Metals having this combination of properties are used in vital structural members that may become overloaded in service.

In a discussion of metal properties, stress and strain should be mentioned. Stress is a force placed upon a body and is measured in terms of force per unit area, the force being expressed in pounds and the unit of area in square inches; in other words, pounds per square inch (psi). Stress may be in the form of compression, tension, torsion, bending, shearing loads or a combination of two or more of these. All parts of an aircraft are subjected to stresses. When a part fails to return to its original form after being stressed, it is said to be strained. The various stresses acting on parts of an aircraft, while in flight, have an important bearing on the metals used:

• Tension. Tension is the resistance to pulling apart or stretching produced by two forces pulling in opposite directions along the same straight line. The cables of a control system are placed under tension when the controls are operated.

• Compression. Compression is the resistance to pushing together or crushing produced by two forces pushing toward each other in the same straight line. The landing struts of an aircraft are under compression when landing and, to a smaller degree, when supporting the weight of the aircraft as it rests on the ground.

• Torsion. Torsion is the resistance to twisting. A torsional force is produced when an engine turns a crankshaft. Torque is the force that produces torsion.

• Bending. Bending is a combination of tension and compression. The inside curve of the bent object is under compression, and the outside curve is under tension. The main spars of the main rotor blades are subjected to bending. The blades droop while the rotor head is at rest and bend upward when rotating.

• Shear. Shear is the stress exerted when two pieces of metal fastened together are separated by sliding one over the other in opposite directions. The stress cuts off a bolt or a rivet like a pair of shears. In general, rivets are subjected to shear only; bolts, to both shear and tension. There is internal shear in most bending elements and in the skin of sheet metal structures.

2. Metallurgical and Metalworking Terms

Some of the more commonly used terms in the field of metallurgy are listed below:

• Strength

◇Creep strength—the only ability of a metal to resist slow deformation due to stress, but less than the stress level needed to reach the yield point. Creep strength is usually stated in terms of time, temperature and load.

◇Yield strength—the point reached when metal exhibits a permanent set under load.

◇Rupture strength—the point where metal will break under a continual load applied for periods of 100 and 1,000 hours. Metals are usually tested at several temperatures.

◇Ultimate tensile strength—the load under which metal will break in a short time.

• Ductility—the ability of a metal to deform without breaking.

• Coefficient of expansion—a measure of how much a metal will expand or grow with the application of heat.

• Thermal conductivity—a measure of the ability of a metal to transmit heat.

• Corrosion and oxidation resistance—indicates how well a metal can resist the corrosion effects of the hot exhaust stream.

• Melting point—the temperature at which metal becomes a liquid.

• Critical temperature—point where as it is cooled a metal's internal structure and physical properties are altered. The rate of cooling will greatly influence the ultimate properties of the metal.

• Heat treatability—a measure of how the metal's basic structure will vary under an operation or series of operations involving heating and cooling of the metal while it is in a solid state. Ferritic, austenitic and martensitic steels all vary in heat treatability.

• Thermal shock resistance—the ability of a metal to withstand extreme changes in temperature in short periods of time.

• Hardness—the ability of a metal resist abrasion penetration cutting action or permanent distortion. Hardness may be increased by working the metal or by heat treatment.

• Alloying—the combining of metal in their molten state to create desirable qualities or reduce or eliminate undesirable qualities.

Common metal working terms include the following:

• Casting—a process whereby molten metal solidifies in a mold.

• Forging—a process of plastic deformation under pressure that may be slowly or quickly applied.

- Electrochemical machining (ECM)—controlled high-speed deplating using a shaped tool (cathode), an electricity-conducting solution and the workpiece (anode).

- Machining—any process where metal is formed by cutting, hot or cold rolling, pinching, punching, grinding, or by laser beams.

- Extrusion—pushing through a die to form various cross-sectional shapes.

- Welding—a process of fusing two pieces of metal together by locally melting part of the material by arc welders, plasmas, lasers, or electron beams.

- Pressing—a process of blending, pressing, sintering (fusing the power particles together through heat), and coining metals out of prealloyed powders.

- Protective finish and surface treatment—includes plating by electrical and chemical processes ceramic coatings, or painting. Surface treatments for increased wear may take the form of nitriding, cyaniding, carburizing, diffusion, coating and flame plating.

- Shot peening—a plastic flow or stretching of a metal's surface by a rain of round metallic shot throw at high velocity. Shot peening also serves to work-harden metals, especially aluminum alloy.

- Heat treatment—a series of operations involving the controlled heating and cooling of metals in the solid state. Its purposes is to change a mechanical property or combination of properties so that the metal will be more useful, serviceable, and safe for a specific purpose. It includes normalizing annealing stress relieving, tempering, and hardening.

- Inspection (strictly speaking not a part of the metalworking process but integrally associated with it)—includes magnetic particle and dye penetrant inspection, X-ray inspection dimensional and visual inspection, and inspection by devices using sound, light, and air.

3. Heat Ranges of Metals

The operating conditions inside a gas turbine engine vary considerably, and metals differ in their ability to satisfactorily meet these conditions.

- Aluminum alloys. Aluminum and its alloys are used temperature rangers up to 500 °F. With low density and good strength-to-weight ratios, aluminum forgings and castings are used extensively for centrifugal compressor wheels and housings, air inlet sections, accessory sections, and for the accessories themselves.

- Magnesium alloys. Magnesium is the lightest structural metal in the world. Aluminum is 15 times heavier, titanium 25 times heavier, steel 4 times heavier, and copper and nickel alloys are 5 times heavier. Magnesium is combined with small amounts of certain other metals, including aluminum, manganese, zinc, zirconium, thorium, and others, to obtain the strong lightweight alloys needed for structural purposes.

• Titanium alloys. Titanium and its alloys are used for axial-flow compressor wheels, blades, and other forged components in many large, high-performance engines. Titanium combines high strength with low density and is suitable for applications up to 100°F.

• Steel alloys. This group includes high-chromium, molybdenum, high-nickel, and iron-base alloys in addition alloys in addition to low-alloy steels. Because of the relatively low material cost, ease of fabrication, and good mechanical properties, low-alloy steels are commonly used for both rotating and static engine components such as compressor rotor blades, wheels, spacers, stator vanes and structural members. Low-alloy steels can be heat-treated and can withstand temperatures up to 100°F. High nickel chromium iron-base alloys can be used up to 1,250°F.

• Nickel-Base Alloys. Nickel-base alloys are some of the best metals for use between 1,200°F and 1,800°F. Most contain little or no iron. They develop high-temperature strength by age hardening and characterized by long-time creep-rupture strength as well as high ultimate and yield strength combined with good ductility. Many of these materials, originally developed for turbine bucket applications, are also being used in turbine wheels, shafts, spacers, and other parts. Their use is somewhat restricted because of cost and because of their requirement for critical materials.

• Cobalt-Base Alloys. Cobalt-base alloys form another important group of high-temperature, high-strength, and high-corrosion-resistant metals. They contain little or no iron. These alloys are used in afterburner and other parts of the engine subjected to very high temperatures.

4. Chemical Elements of Alloys

The number of materials used in alloys is large. Some of the most commonly used elements are listed below.

The percentages of elements used partially determines the physical and chemical characteristics of the alloy and its suitability to a particular application. Tempering and other processes determine the rest. Three characteristics that must be considered are:

• High-temperature strength;

• Resistance to oxidation and corrosion;

• Resistance to thermal shock.

High-temperature strength. The most highly stressed parts of the gas turbine engine are the turbine blades and discs. Centrifugal forces tending to break the disc very with the square of the speed. For example, the centrifugal force in a disc rotating at 200,000 RPM will be four times that at 10,000 RPM. Blades weighing only 2 ounces may exert loads of over 4,000 pounds at maximum RPM. Blades must also resist the high bending loads applied by the moving gas stream to produce the thousands of horsepower (difference) of several hundred degrees between the central portion of the disk and its periphery.

Many metals which would be quite satisfactory at room temperatures will lose much of their strength at the elevated temperatures encountered in the engine's hot section. The ultimate ensile strength of a metal at one temperature dose not necessarily indicate its ultimate tensile strength at a higher temperature. For example, at 1,000°F. Inconel X has an ultimate tensile strength of approximately 160,000 psi; and S 816 at the same temperature has an ultimate tensile strength of 135,000 psi. At 1,500°F their positions are reversed. Inconel X has an ultimate tensile strength of 55,000 psi; S 816 has an ultimate tensile strength of 75,000 psi. The creep strength, which is closely associated with untimate tensile strength, is probably one of the most important considerations in the selection of a suitable metal for turbine blades, engine vibration and fatigue resistance will also have some influence on the selection and useful life of both discs and blades.

Many materials will withstand the high temperatures encountered in a gas turbine engine (carbon columbium, molybdenum, rhenium tantalum, and tungsten all have melting points above 4,000°F). However, the ability of withstand high temperatures while maintaining reasonable tensile strength is not the only consideration. All of the following qualities must be taken into account when selecting a particular metal:

- Critical temperature;
- Rupture strength;
- Thermal conductivity;
- Coefficient of expansion;
- Yield strength;
- Ultimate tensile strength;
- Corrosion resistance;
- Workability;
- Cost.

Resistance to Oxidation and Corrosion. Corrosion and oxidation are results of electrical and chemical reactions with other materials. The hot exhaust gas stream encountered in the engine speeds up the reaction. While all metals will corrode or oxidize, the degree of oxidation is determined by the base alloy and the properties of the oxide coating formed. If the oxide coating is porous or has a coefficient of expansion different from that of the base metal, the base metal will be continually exposed to the oxidizing atmosphere. One solution to oxidation at elevated temperatures is ceramic coatings. Ceramic coated afterburner liners and combustion chambers are in use today. The ceramic coating has two basic functions:

- Sealing the base metal surface against corrosion;
- Insulating the base metal against high temperatures.

These coatings are not without disadvantages:

- They are more susceptible to thermal shock;
- They must have the same coefficient of expansion as the base metal;
- They are brittle;
- They have low tensile strength which restricts their use in the engine.

Some promising work is being done with various metal ceramic combinations called cermets or caramels. Materials being used with ceramics include aluminum, beryllium, thorium and zirconium oxides, to name a few, resistance to thermal shock. Many materials which would otherwise be quite suitable must be rejected because of their poor thermal shock characteristics, several engine failures have been attributed to thermal shock on the turbine disc. Ceramic coating in particular are vulnerable to this form of stress. Improved fuel controls, starting techniques and engine design have lessened this problem.

5. Transpiration Cooling

The effort to achieve higher turbine inlet temperatures (and therefore higher thermal efficiency) has been approached from two directions: high temperature materials and cooling methods. A common method of cooling the nozzle guide vanes on gas turbine engines is to pass compressor bleed air through the hollow blades to cool them by connective heat transfer some engines also use air bled from the compressor to cool the front and rear face of the turbine discs and the hollow turbine blades.

Transpiration cooling is a novel and efficient method of allowing the turbine blades and other parts within the hot section to operate at much higher turbine inlet temperatures. The Wright Corporation has constructed and run turbine blades at an inlet temperature of 2,500°F. In this type of cooled blade the air passes through thousands of holes in a porous airfoil made from a sintered wire mesh material. Since the sintered wire mesh is not strong enough by itself, an internal strut is provided as the main structural support carrying all air foil and centrifugal loads.

Fabrication techniques involve rolling layers of woven wire mesh and then sintering these layers to forma porous metal sheet. The sheet is then rolled into an airfoil shape.

Porous materials have been tested for use in combustion chambers and for afterburner liners. A similar called Rigimesh has also been used in rocket engines to help keep the fuel nozzles cool. Many manufacturers are experimenting with other types of porous materials for use in blades in an attempt to obtain higher turbine inlet temperatures.

6. Other Materials

Relatively new materials called composites are coming into use in both airframes and engines. In these products graphite, glass or boron filaments are embedded in an epoxy-resin matrix or base substance. Other types of filaments and materials are being tried to meet the demands of higher temperatures and stress the chief advantage of composite material is its very favorable strength-to-weight ratio, which can lead to lighter weight in many structural parts. For example, a lighter fan blade allows a lighter fan disc, which in turn permits a lightening of other parts all the way down the line. Composite materials mat be used in conjunction with other load-bearing materials to provide a support function. Typical of this type of structure are fan blades with a steel spar and base and base and with an airfoil composite shell.

Vocabulary

material	/məˈtɪərɪəl/	n. 衣料，布料；材料，原料
unique	/juˈniːk/	adj. 独一无二的，独特的；非常特别的，极不寻常的
limit	/ˈlɪmɪt/	n. 限度，限制；极限，限量；（区域的）界限，边界
correspond	/ˌkɒrəˈspɒnd/	v. 类似于，相当于；通信；相一致，符合
fuel consumption		耗油量
properties	/ˈprɒpətiz/	n. 属性
metallurgist	/məˈtælədʒɪst/	n. 冶金学者；冶金家
be familiar with		熟悉
hardness	/ˈhɑːdnəs/	n. /物/ 硬度；坚硬；困难；冷酷
brittleness	/ˈbrɪtlnəs/	n. /材/ 脆性，/材/ 脆度；脆弱性
malleability	/ˌmælɪəˈbɪləti/	n. 顺从；可锻性；展延性
ductility	/dʌkˈtɪlɪti/	n. 延展性；柔软性；顺从
toughness	/ˈtʌfnəs/	n. 韧性
conductivity	/ˌkɒndʌkˈtɪvəti/	n. 导电性；/物//生理/ 传导性
strength	/streŋkθ/	n. 强度
bending	/ˈbendɪŋ/	n. 弯曲，弯曲度
shear	/ʃɪə(r)/	n. 剪应力

Exercise

Complete the following sentences and translate them into Chinese.

(1) High-temperature, high-strength materials and unique methods of manufacture have made the gas turbine engine_____.

翻译：_____

_____

(2) Each type of metal or alloy has_____which make it desirable for a particular use, but it may have other qualities which are undesirable.

翻译：_____

_____

(3) Hardness refers to _____, penetration, cutting action or permanent distortion.

翻译：_____

_____

(4) Brittleness is_____which allows little bending or deformation without shattering.

翻译：_____

_____

(5) Ductility is the property of a metal which permits it to be permanently drawn, bent, or twisted into_____.

翻译：_____

_____

(6) Elasticity is that property which _____when the force which causes the change of shape is removed.

翻译：_____

_____

(7) A material_____will with stand tearing or shearing and may be stretched or otherwise deformed without breaking.

翻译：_____

_____

(8) Conductivity is the property which enables a metal to carry_____.

翻译：_____

_____

(9) Strength refers to the ability of a metal to＿＿＿＿＿＿＿＿＿＿＿＿＿. Strength is a property that sums up many of the desirable qualities of metals.

翻译：＿＿＿＿＿＿＿＿＿＿＿＿＿＿＿＿＿＿＿＿＿＿＿＿＿＿＿＿＿＿＿＿＿＿＿＿

＿＿＿＿＿＿＿＿＿＿＿＿＿＿＿＿＿＿＿＿＿＿＿＿＿＿＿＿＿＿＿＿＿＿＿＿＿＿＿

(10) Shear is the stress exerted when＿＿＿＿＿＿＿＿＿＿＿＿＿ are separated by sliding one over the other in opposite directions.

翻译：＿＿＿＿＿＿＿＿＿＿＿＿＿＿＿＿＿＿＿＿＿＿＿＿＿＿＿＿＿＿＿＿＿＿＿＿

＿＿＿＿＿＿＿＿＿＿＿＿＿＿＿＿＿＿＿＿＿＿＿＿＿＿＿＿＿＿＿＿＿＿＿＿＿＿＿

Supplementary Reading

Future of Propulsion

There is continuing research into the design of more efficient engines that use less fuel and generate higher thrust per weight. New advances in the study of high temperature materials are allowing the design of turbines that can tolerate even higher inlet temperatures. Advances in aerodynamics calculations and experiments are leading to improvements in the designs of the inlets, compressors, turbines and nozzles. And a relatively new area, active control theory, is showing promising results. In an active control system, computer chips monitor the conditions throughout the engine to determine the optimum component behavior for that particular flight. The computer may change the fuel and air ratio in the combustor slightly for a better burn, or it might change the shape of either the inlet or the exit nozzle slightly to improve the aerodynamics.

In supersonic and hypersonic engine design, work continues to find better solutions for engines. The challenge here is particularly daunting; the final choices must work across a wide range of flight mach numbers and conditions. Hybrid configurations that use combinations of the turbojet, ramjet and rocket engines are the focus of study these days, and many hybrid engines are in the conceptual-design stages.

Whether the goal is to improve existing engines for the world's transportation needs, or designing the engines for future supersonic and hypersonic aircraft, there is plenty of room for more engineers and more ideas. There are many challenges out there!

参 考 文 献

[1] 空客 A320 飞机维护手册(Aircraft Maintenance Manual). 2005.

[2] 波音 747 - 400 飞机维护手册(Aircraft Maintenance Manual). 2002.

[3] 波音 737 - 300 飞机维护手册(Aircraft Maintenance Manual). 2001.

[4] Aviation Maintenance Technician Handbook-Powerplant. 2012.

[5] 李书明. 航空器动力装置修理. 北京:中国科学文化出版社,2003.

[6] 刘长福,邓明. 航空发动机结构分析. 西安:西北工业大学出版社,2006.

[7] 赵迎春,陈凯军. 飞机维修专业英语. 北京:中国水利水电出版社,2018.

[8] 邓明. 航空燃气涡轮发动机原理与构造. 北京:国防工业出版社,2008.

[9]《航空发动机设计手册》编委会. 航空发动机设计手册. 北京:航空工业出版社,2000.

[10] 李永平. 民航机务专业英语 3. 北京:清华大学出版社,2018.

[11] 刘志武. 航空专业英语. 北京:北京理工大学出版社,2010.